W9-APE-978

An Introduction to Genetic Engineering

Des Nicholl presents here a new, fully revised, and expanded edition of his popular undergraduate-level textbook. Many of the features of the original edition have been retained; the book still offers a concise technical introduction to the subject of genetic engineering. However, the book is now divided into three main sections: the first introduces students to basic molecular biology, the second section explains the methods used to manipulate genes, and the third deals with modern applications of genetic engineering. A whole chapter is now devoted to the polymerase chain reaction. Applications covered in the book include genomics, protein engineering, gene therapy, cloning, and transgenic animals and plants. A final chapter discusses the ethical questions surrounding genetic engineering in general.

An Introduction to Genetic Engineering is essential reading for undergraduate students of biotechnology, genetics, molecular biology and biochemistry.

DES NICHOLL is a Senior Lecturer in Biological Sciences, The University of Paisley, Scotland.

Studies in Biology series is published in association with the Institute of Biology (London, UK). The series provides short, affordable and very readable textbooks, aimed primarily at undergraduate biology students. Each book offers either an introduction to a broad area of biology (e.g. *Introductory Microbiology*), or a more in-depth treatment of a particular system or specific topic (e.g. *Photosynthesis*). All of the subjects and systems covered are selected on the basis that all undergraduate students will study them at some point during their biology degree courses.

Titles available in this series

Photosynthesis, 6th edition, D. O. Hall and K. K. Rao

Introductory Microbiology, J. Heritage, E. G. V. Evans and R. A. Killington

Biotechnology, 3rd edition, J. E. Smith

An Introduction to Parasitology, B. E. Matthews

Essentials of Animal Behaviour, P. J. B. Slater

Microbiology in Action, J. Heritage, E. G. V. Evans and R. A. Killington

An Introduction to the Invertebrates, J. Moore

Nerve and Muscle, 3rd edition, R. D. Keynes and D. J. Aidley

An Introduction to Genetic Engineering, 2nd edition, D. S. T. Nicholl

An Introduction to Genetic Engineering

Second edition

Desmond S. T. Nicholl
Senior Lecturer in Biological Sciences, University of Paisley

PUBLISHED BY THE PRESS SYNDICATE OF THE UNIVERSITY OF CAMBRIDGE
The Pitt Building, Trumpington Street, Cambridge, United Kingdom

CAMBRIDGE UNIVERSITY PRESS
The Edinburgh Building, Cambridge CB2 2RU, UK
40 West 20th Street, New York, NY 10011-4211, USA
477 Williamstown Road, Port Melbourne, VIC 3207, Australia
Ruiz de Alarcón 13, 28014 Madrid, Spain
Dock House, The Waterfront, Cape Town 8001, South Africa

http://www.cambridge.org

First published 1994
Second edition 2002

Printed in the United Kingdom at the University Press, Cambridge

Typeface Monotype Garamond 11/13pt *System* QuarkXPress™ [SE]

A catalogue record for this book is available from the British Library

Library of Congress Cataloguing in Publication data

Nicholl, Desmond S. T.
An introduction to genetic engineering/Desmond S. T. Nicholl.
p. cm.
Previous edition published in the series: Studies in biology.
Includes bibliographical references and index.
ISBN 0 521 80867 7 (hardback) – ISBN 0 521 00471 3 (pbk.)
1. Genetic engineering. I. Title.

QH442.N53 2002
660.65–dc21 2001037542

ISBN 0 521 80867 7 hardback
ISBN 0 521 00471 3 paperback

Contents

Preface to the second edition

Advances in genetics continue to be made at an ever increasing rate, which makes writing an introductory text somewhat difficult. In the few years since the first edition was published, many new applications of gene manipulation technology have been developed, covering a diverse range of disciplines. The temptation in preparing this second edition was to concentrate on the applications, and ignore the fundamental principles of the technology. However, I wished to retain many of the features of the first edition, in which a basic technical introduction to the subject was the main aim of the text. Thus some of the original methods used in gene manipulation have been kept as examples of how the technology developed, even though some of these have become little used or even obsolete. From the educational point of view, this should help the reader cope with more advanced information about the subject – a sound grasp of the basic principles is an important part of any introduction to genetic engineering. I have been gratified by the many positive comments about the first edition, and I hope that this new edition is as well received.

In trying to strike a balance between the methodology and the applications of gene manipulation, I have divided the text into three sections. Part I deals with basic molecular biology, Part II with the methods used to manipulate genes, and Part III with the applications. These sections may be taken out of order if desired, depending on the level of background knowledge. Apart from a general revision of chapters retained from the first edition, there have been some more extensive changes made. The increasing importance of the polymerase chain reaction is recognised by a new chapter devoted to this topic. In Part III there are now five separate chapters dealing with the applications of gene manipulation, as opposed to a single chapter in the first edition. I hope that the changes have produced a balanced treatment of the field, whilst retaining the introductory nature of the text and keeping it to a reasonable length.

My thanks go to my colleagues Simon Hettle, John McLean, Ros Brett and Anne Dickson for comments on various parts of the manuscript. Their help has made the book better; any errors of fact or interpretation remain my own responsibility. My final and biggest thank you goes to my wife Linda and to Charlotte, Thomas and Anna. They have suffered with me during the writing, and have put up with more than they should have had to. I dedicate this edition to them.

Desmond S. T. Nicholl
Paisley

1

Introduction

1.1 What is genetic engineering?

Progress in any scientific discipline is dependent on the availability of techniques and methods that extend the range and sophistication of experiments which may be performed. Over the last 30 years or so this has been demonstrated in a spectacular way by the emergence of genetic engineering. This field has grown rapidly to the point where, in many laboratories around the world, it is now routine practice to isolate a specific DNA fragment from the genome of an organism, determine its base sequence, and assess its function. The technology is also now used in many other applications, including forensic analysis of scene-of-crime samples, paternity disputes, medical diagnosis, genome mapping and sequencing, and the biotechnology industry. What is particularly striking about the technology of gene manipulation is that it is readily accessible by individual scientists, without the need for large-scale equipment or resources outside the scope of a reasonably well-found research laboratory.

The term **genetic engineering** is often thought to be rather emotive or even trivial, yet it is probably the label that most people would recognise. However, there are several other terms that can be used to describe the technology, including **gene manipulation**, **gene cloning**, **recombinant DNA technology**, **genetic modification**, and the **new genetics**. There are also legal definitions used in administering regulatory mechanisms in countries where genetic engineering is practised.

Although there are many diverse and complex techniques involved, the basic principles of genetic manipulation are reasonably simple. The premise

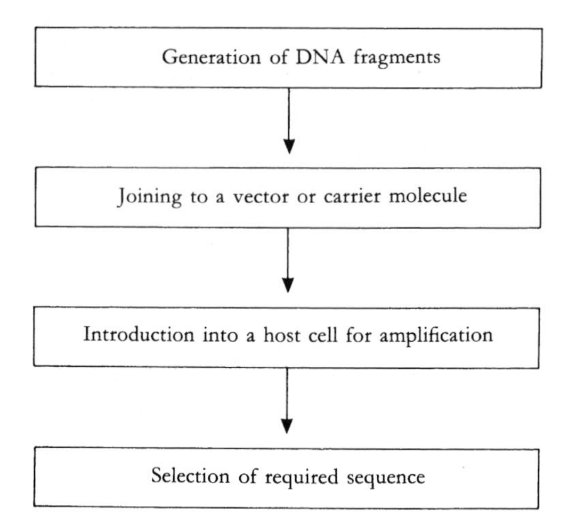

Fig. 1.1. The four steps in a gene cloning experiment. The term *clone* comes from the colonies of identical host cells produced during amplification of the cloned fragments. Gene cloning is sometimes referred to as *molecular cloning*, to distinguish the process from the cloning of whole organisms.

on which the technology is based is that genetic information, encoded by DNA and arranged in the form of genes, is a resource that can be manipulated in various ways to achieve certain goals in both pure and applied science and medicine. There are many areas in which genetic manipulation is of value, including:

- Basic research on gene structure and function
- Production of useful proteins by novel methods
- Generation of transgenic plants and animals
- Medical diagnosis and treatment.

In later chapters I look at some of the ways in which genetic manipulation has contributed to these areas.

The mainstay of genetic manipulation is the ability to isolate a single DNA sequence from the genome. This is the essence of **gene cloning**, and can be considered as a series of four steps (Fig. 1.1). Successful completion of these steps provides the genetic engineer with a specific DNA sequence, which may then be used for a variety of purposes. A useful analogy is to consider gene cloning as a form of **molecular agriculture**, enabling the production of large amounts (in genetic engineering this means micrograms or milligrams) of a particular DNA sequence.

One aspect of the new genetics that has given cause for concern is the debate surrounding the potential applications of the technology. The term **genethics** has been coined to describe the ethical problems that exist in modern genetics, which are likely to increase in both number and complexity as genetic engineering technology becomes more sophisticated. The use of transgenic plants and animals, investigation of the human genome, gene therapy, and many other topics are of concern not just to the scientist but to the population as a whole. The recent developments in genetically modified foods have provoked a public backlash against the technology. Additional developments in the cloning of organisms, and in areas such as *in vitro* fertilisation and xenotransplantation, raise further questions. Although not strictly part of gene manipulation technology, I will consider aspects of organismal cloning later in this book, as this is an area of much concern and can be considered as genetic engineering in its broadest sense.

Taking all the potential costs and benefits into account, it remains to be seen if we can use genetic engineering for the overall benefit of mankind, and avoid the misuse of technology that often accompanies scientific achievement.

1.2 Laying the foundations

Although the techniques used in gene manipulation are relatively new, it should be remembered that development of these techniques was dependent on the knowledge and expertise provided by microbial geneticists. We can consider the development of genetics as falling into three main eras (Fig. 1.2). The science of genetics really began with the rediscovery of Gregor Mendel's work at the turn of the century, and the next 40 years or so saw the elucidation of the principles of inheritance and genetic mapping. Microbial genetics became established in the mid-1940s, and the role of DNA as the genetic material was confirmed. During this period great advances were made in understanding the mechanisms of gene transfer between bacteria, and a broad knowledge base was established from which later developments would emerge.

The discovery of the structure of DNA by James Watson and Francis Crick in 1953 provided the stimulus for the development of genetics at the molecular level, and the next few years saw a period of intense activity and excitement as the main features of the gene and its expression were determined. This work culminated with the establishment of the complete genetic code in 1966 – the stage was now set for the appearance of the new genetics.

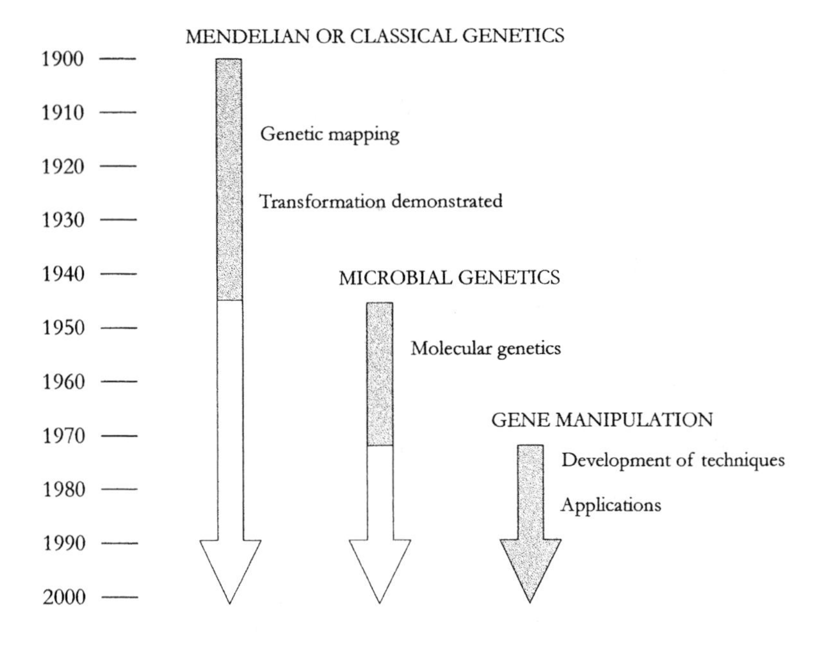

Fig. 1.2. The history of genetics since 1900. Shaded areas represent the periods of major development in each branch of the subject.

1.3 First steps

In the late 1960s there was a sense of frustration among scientists working in the field of molecular biology. Research had developed to the point where progress was being hampered by technical constraints, as the elegant experiments that had helped to decipher the genetic code could not be extended to investigate the gene in more detail. However, a number of developments provided the necessary stimulus for gene manipulation to become a reality. In 1967 the enzyme **DNA ligase** was isolated. This enzyme can join two strands of DNA together, a prerequisite for the construction of recombinant molecules, and can be regarded as a sort of molecular glue. This was followed by the isolation of the first **restriction enzyme** in 1970, a major milestone in the development of genetic engineering. Restriction enzymes are essentially molecular scissors, which cut DNA at precisely defined sequences. Such enzymes can be used to produce fragments of DNA that are suitable for joining to other fragments. Thus, by 1970, the basic tools required for the construction of recombinant DNA were available.

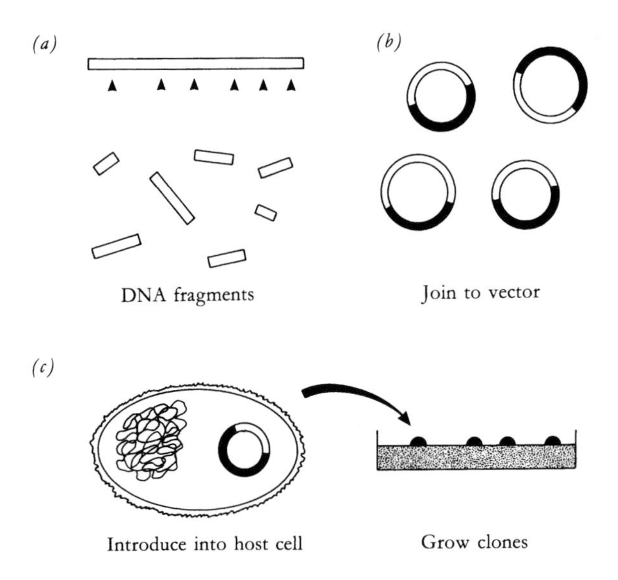

Fig. 1.3. Cloning DNA fragments. (*a*) The source DNA is isolated and fragmented into suitably sized pieces. (*b*) The fragments are then joined to a carrier molecule or vector to produce recombinant DNA molecules. In this case, a plasmid vector is shown. (*c*) The recombinant DNA molecules are then introduced into a host cell (a bacterial cell in this example) for propagation as clones.

The first recombinant DNA molecules were generated at Stanford University in 1972, utilizing the cleavage properties of restriction enzymes (scissors) and the ability of DNA ligase to join DNA strands together (glue). The importance of these first tentative experiments cannot be overestimated. Scientists could now join different DNA molecules together, and could link the DNA of one organism to that of a completely different organism. The methodology was extended in 1973 by joining DNA fragments to the plasmid pSC101, which is an **extrachromosomal element** isolated from the bacterium *Escherichia coli*. These recombinant molecules behaved as **replicons**, i.e. they could replicate when introduced into *E. coli* cells. Thus, by creating recombinant molecules *in vitro*, and placing the construct in a bacterial cell where it could replicate *in vivo*, specific fragments of DNA could be isolated from bacterial colonies that formed clones (colonies formed from a single cell, in which all cells are identical) when grown on agar plates. This development marked the emergence of the technology which became known as **gene cloning** (Fig. 1.3).

The discoveries of 1972 and 1973 triggered off what is perhaps the biggest scientific revolution of all – the new genetics. The use of the new technology spread very quickly, and a sense of urgency and excitement prevailed. This was dampened somewhat by the realisation that the new technology could give rise to potentially harmful organisms, exhibiting undesirable characteristics. It is to the credit of the biological community that measures were adopted to regulate the use of gene manipulation, and that progress in contentious areas was limited until more information became available regarding the possible consequences of the inadvertent release of organisms containing recombinant DNA. However, the development of genetically modified organisms (GMOs), particularly crop plants, has re-opened the debate about the safety of these organisms and the consequences of releasing GMOs to the environment.

1.4 What is in store?

In writing a second edition of this book, I have reorganised much of the content to better reflect the current applications of DNA technology. However, I have retained introductory material on molecular biology, working with nucleic acids, and on the basic methodology of gene manipulation. I hope that this edition will therefore continue to serve as a technical introduction to the subject, whilst also giving a much broader appreciation of the applications of this exciting range of technologies.

I have organised the remaining chapters into three sections.

Part I (*The basis of genetic engineering*; Chapters 2–4) deals with the basic techniques. Chapter 2 (*Introducing molecular biology*) and Chapter 3 (*Working with nucleic acids*) provide background information about DNA and the techniques used when working with it. Chapter 4 (*The tools of the trade*) looks at the range of enzymes needed for gene manipulation.

Part II (*The methodology of gene manipulation*; Chapters 5–8) outlines the techniques and strategies needed to clone and identify genes. Chapter 5 (*Host cells and vectors*) and Chapter 6 (*Cloning strategies*) describe the various systems and protocols that may be used to clone DNA. Chapter 7 is dedicated to the *Polymerase chain reaction*, which has revolutionised many areas of molecular biology. Chapter 8 (*Selection, screening and analysis of recombinants*) describes how particular DNA sequences can be selected from collections of cloned fragments.

In **Part III** (*Genetic engineering in action*; Chapters 9–14) the applications of gene manipulation and associated technologies are discussed. Topics covered are *Understanding genes and genomes* (Chapter 9), *Genetic engineering and biotechnology* (Chapter 10), *Medical and forensic applications* (Chapter 11), and *Transgenic plants* *and animals* (Chapter 12). Organismal cloning is examined in Chapter 13 (*The other sort of cloning*), and the moral and ethical considerations of genetic engineering are considered in Chapter 14 (*Brave new world or genetic nightmare?*).

At the end of each chapter a **concept map** is given, covering the main points of the chapter. Concept mapping is a technique that can be used to structure information and provide links between various topics. The concept maps provided here are essentially summaries of the chapters, and may be examined either before or after reading the chapter. I hope that they prove to be a useful addition to the text.

Suggestions for further reading are given at the end of the book, along with tips for using the internet and world wide web. No reference has been made to the primary literature, as this is well documented in the books and articles mentioned in this section. A glossary of terms used has also been provided; this may be particularly useful for readers who may be unfamiliar with the terminology used in molecular biology.

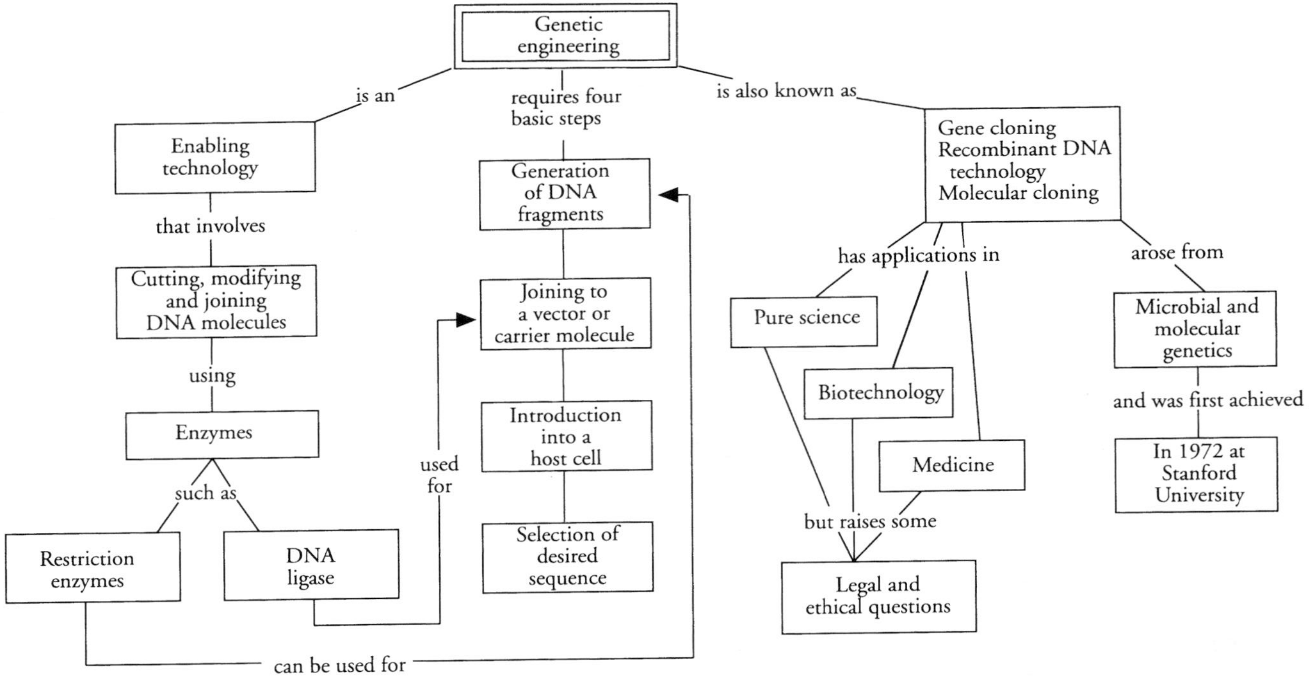

Genetic engineering
is an
requires four basic steps
is also known as
Enabling technology
that involves
Cutting, modifying and joining DNA molecules
using
Enzymes
such as
Restriction enzymes
DNA ligase
used for
can be used for
Generation of DNA fragments
Joining to a vector or carrier molecule
Introduction into a host cell
Selection of desired sequence
Gene cloning
Recombinant DNA technology
Molecular cloning
has applications in
Pure science
Biotechnology
Medicine
but raises some
Legal and ethical questions
arose from
Microbial and molecular genetics
and was first achieved
In 1972 at Stanford University

Concept map 1

Part I

The basis of genetic engineering

2

Introducing molecular biology

In this chapter I present a brief overview of the structure and function of DNA, and its organisation within the genome (the total genetic complement of an organism). This provides the non-specialist reader with an introduction to the topic, and may also act as a useful refresher for those who have some background knowledge of DNA. More extensive accounts of the topics presented here may be found in the textbooks listed in Suggestions for further reading.

2.1 The flow of genetic information

It is a remarkable fact that an organism's characteristics are encoded by a four-letter alphabet, defining a language of three-letter words. The letters of this alphabet are the bases adenine (A), guanine (G), cytosine (C) and thymine (T), with triplet combinations of these bases making up the 'dictionary' that is the genetic code.

The expression of genetic information is achieved ultimately *via* proteins, particularly the **enzymes** that catalyse the reactions of metabolism. Proteins are condensation **heteropolymers** synthesised from amino acids, of which 20 are used in natural proteins. Given that a protein may consist of several hundred amino acid residues, the number of different proteins that may be made is essentially unlimited; thus great diversity of protein form and function can be achieved using an elegantly simple coding system. The genetic code is shown in Table 2.1.

The flow of genetic information is unidirectional, from DNA to protein, with **messenger RNA** (mRNA) as an intermediate. The copying of

Table 2.1. *The genetic code*

First base (5′-end)	Second base				Third base (3′-end)
	U	C	A	G	
U	Phe	Ser	Tyr	Cys	U
	Phe	Ser	Tyr	Cys	C
	Leu	Ser	STOP	STOP	A
	Leu	Ser	STOP	Trp	G
C	Leu	Pro	His	Arg	U
	Leu	Pro	His	Arg	C
	Leu	Pro	Gln	Arg	A
	Leu	Pro	Gln	Arg	G
A	Ile	Thr	Asn	Ser	U
	Ile	Thr	Asn	Ser	C
	Ile	Thr	Lys	Arg	A
	Met	Thr	Lys	Arg	G
G	Val	Ala	Asp	Gly	U
	Val	Ala	Asp	Gly	C
	Val	Ala	Glu	Gly	A
	Val	Ala	Glu	Gly	G

Note: Uracil (U) is found in RNA in place of thymine, and is in this table. Codons read 5′ 3′, thus AUG specifies Met. The three-letter abbreviations for the amino acids are as follows: Ala, Alanine; Arg, Arginine; Asn, Asparagine; Asp, Aspartic acid; Cys, Cysteine; Gln, Glutamine; Glu, Glutamic acid; Gly, Glycine; His, Histidine; Ile, Isoleucine; Leu, Leucine; Lys, Lysine; Met, Methionine; Phe, Phenylalanine; Pro, Proline; Ser, Serine; Thr, Threonine; Trp, Tryptophan; Tyr, Tyrosine; Val, Valine. The three codons UAA, UAG and UGA specify no amino acid and terminate translation (STOP).

DNA-encoded genetic information into RNA is known as **transcription** (T_C), with the further conversion into protein being termed **translation** (T_L). This concept of information flow is known as the **Central Dogma** of molecular biology, and is an underlying theme in all studies on gene expression.

A further two aspects of information flow may be added to this basic model to complete the picture. Firstly, duplication of the genetic material prior to cell division represents a DNA–DNA transfer, and is known as DNA **replication**. A second addition, with important consequences for the genetic engineer, stems from the fact that some viruses have RNA instead of DNA as their genetic material. These viruses (chiefly members of the retrovirus

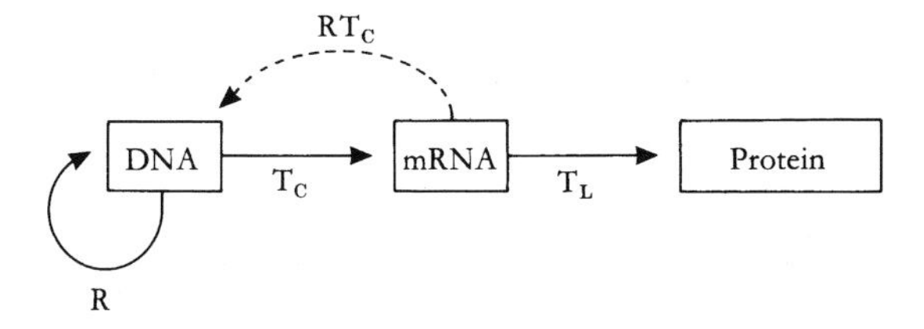

Fig. 2.1. The Central Dogma. This states that information flow is unidirectional, from DNA to mRNA to protein. The processes of transcription (T_C), translation (T_L) and DNA replication (R) obey this rule. An exception is found in some RNA viruses, which carry out a process known as reverse transcription (RT_C), producing a DNA copy of their viral RNA genome.

group) have an enzyme called **reverse transcriptase** (an RNA-dependent DNA polymerase) which produces a double-stranded DNA molecule from the single-stranded RNA genome. Thus in these cases the flow of genetic information is reversed with respect to the normal convention. The Central Dogma is summarised in Fig. 2.1.

2.2 The structure of DNA and RNA

In most organisms, the primary genetic material is double-stranded DNA. What is required of this molecule? Firstly, it has to be stable, as genetic information may need to function in a living organism for up to 100 years or more. Secondly, the molecule must be capable of replication, to permit dissemination of genetic information as new cells are formed during growth and development. Thirdly, there should be the potential for limited alteration to the genetic material (**mutation**), to enable evolutionary pressures to exert their effects. The DNA molecule fulfils these criteria of stability, replicability and mutability, and when considered with RNA provides an excellent example of the premise that 'structure determines function'.

Nucleic acids are heteropolymers composed of monomers known as **nucleotides**; a nucleic acid chain is therefore often called a **polynucleotide**. The monomers are themselves made up of three components: a sugar, a phosphate group, and a nitrogenous base. The two types of nucleic acid (DNA and RNA) are named according to the sugar component of the nucleotide, with DNA having 2′-deoxyribose as the sugar (hence **D**eoxyribo**N**ucleic**A**cid) and RNA having ribose (hence **R**ibo**N**ucleic**A**cid). The sugar/phosphate components of a nucleotide are important in determining the structural characteristics of polynucleotides, with the nitrogenous bases determining their

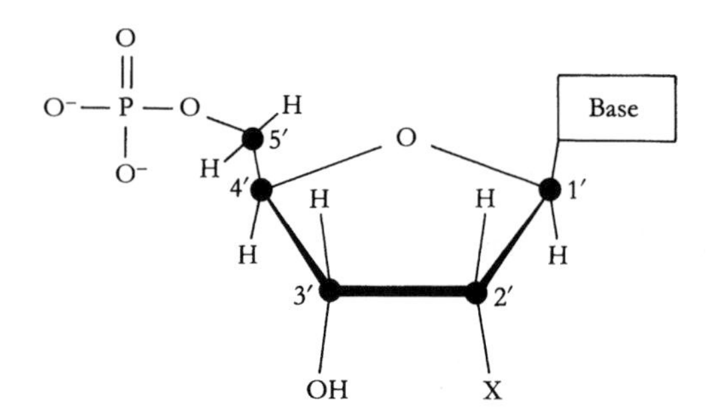

Fig. 2.2. The structure of a nucleotide. Carbon atoms are represented by solid circles, numbered 1′ to 5′. In DNA the sugar is deoxyribose, with a hydrogen atom at position X. In RNA the sugar is ribose, which has a hydroxyl group at position X. The base can be A,G,C or T in DNA, and A,G,C or U in RNA.

information storage and transmission characteristics. The structure of a nucleotide is summarised in Fig. 2.2.

Nucleotides can be joined together by a 5′–3′ phosphodiester linkage, which confers directionality on the polynucleotide. Thus the 5′ end of the molecule will have a free phosphate group, and the 3′ end a free hydroxyl group; this has important consequences for the structure, function and manipulation of nucleic acids. In a double-stranded molecule such as DNA, the sugar–phosphate chains are found in an **antiparallel** arrangement, with the two strands running in different directions.

The nitrogenous bases are the important components of nucleic acids in terms of their coding function. In DNA the bases are as listed in Section 2.1 above, namely adenine (A), guanine (G), cytosine (C) and thymine (T). In RNA the base thymine is replaced by uracil (U), which is functionally equivalent. Chemically adenine and guanine are **purines**, which have a double ring structure, whereas cytosine and thymine (and uracil) are **pyrimidines**, which have a single ring structure. In DNA the bases are paired, A with T and G with C. This pairing is determined both by the bonding arrangements of the atoms in the bases, and by the spatial constraints of the DNA molecule, the only satisfactory arrangement being a purine:pyrimidine base-pair. The bases are held together by hydrogen bonds, two in the case of an A·T base-pair and three in the case of a G·C base-pair. The structure and base-pairing arrangement of the four DNA bases is shown in Fig. 2.3.

The DNA molecule *in vivo* usually exists as a right-handed double helix called the *B*-form. This is the structure proposed by Watson and Crick in 1953. Alternative forms of DNA include the *A*-form (right-handed helix) and the *Z*-form (left-handed helix). Although DNA structure is a complex topic, par-

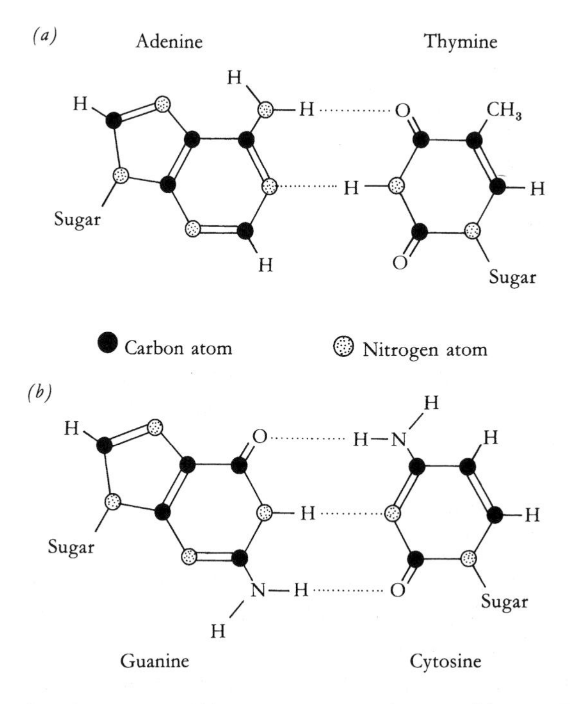

Fig. 2.3. Base-pairing arrangements in DNA. (*a*) An A·T base-pair is shown. The bases are linked by two hydrogen bonds (dotted lines). (*b*) A G·C base-pair, with three hydrogen bonds.

ticularly when the higher-order arrangements of DNA are considered, a simple representation will suffice here, as shown in Fig. 2.4.

The structure of RNA is similar to that of DNA, the main chemical differences being the presence of ribose instead of 2′-deoxyribose and uracil instead of thymine. RNA is also most commonly single-stranded, although short stretches of double-stranded RNA may be found in self-complementary regions. There are three main types of RNA molecule found in cells: **messenger** RNA (mRNA), **ribosomal** RNA (rRNA) and **transfer** RNA (tRNA). Ribosomal RNA is the most abundant class of RNA molecule, making up some 85% of total cellular RNA. It is associated with **ribosomes**, which are an essential part of the translational machinery. Transfer RNAs make up about 10% of total RNA, and provide the essential specificity that enables the insertion of the correct amino acid into the protein that is being synthesised. Messenger RNA, as the name suggests, acts as the carrier of genetic information from the DNA to the translational machinery, and is usually less than 5% of total cellular RNA.

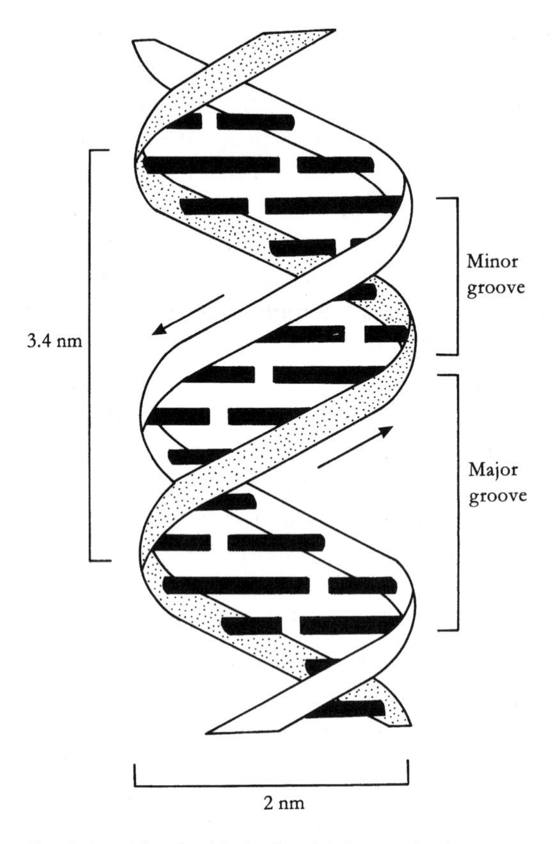

Fig. 2.4. The double helix. This is DNA in the commonly found *B*-form. The right-handed helix has a diameter of 2 nm and a pitch of 3.4 nm, with 10 base-pairs per turn. The sugar–phosphate 'backbones' are antiparallel (arrowed) with respect to their 5′ 3′ orientations. One of the sugar–phosphate chains has been shaded for clarity. The purine–pyrimidine base-pairs are formed across the axis of the helix.

2.3 Gene organisation

The gene can be considered as the basic unit of genetic information. Genes have been studied since the turn of the century, when genetics became established. Before the advent of molecular biology and the realisation that genes were made of DNA, study of the gene was largely indirect; the effects of genes were observed in phenotypes and the 'behaviour' of genes was analysed. Despite the apparent limitations of this approach, a vast amount of information about how genes functioned was obtained, and the basic tenets of transmission genetics were formulated.

As the gene was studied in greater detail, the terminology associated with this area of genetics became more extensive, and the ideas about genes were modified to take account of developments. The term **gene** is usually taken to represent the genetic information transcribed into a single RNA molecule, which is in turn translated into a single protein. Exceptions are genes for RNA molecules (such as rRNA and tRNA), which are not translated. Genes are located on **chromosomes**, and the region of the chromosome where a particular gene is found is called the **locus** of that gene. In diploid organisms, which have their chromosomes arranged as homologous pairs, different forms of the same gene are known as **alleles**.

The double-stranded DNA molecule has the potential to store genetic information in either strand, although in most organisms only one strand is used to encode any particular gene. There is the potential for confusion with the nomenclature of the two DNA strands, which may be called coding/non-coding, sense/antisense, plus/minus, transcribed/non-transcribed or template/non-template. In some cases different authors use the same terms in different ways, which adds to the confusion. Current recommendations from the International Union of Biochemistry (IUB) and the International Union of Pure and Applied Chemistry (IUPAC) favour the terms coding/non-coding, with the **coding** strand of DNA taken to be the **mRNA-like** strand. This convention will be used in this book where coding function is specified. The terms **template** and **non-template** will be used to describe DNA strands when there is not necessarily any coding function involved, as in the copying of DNA strands during cloning procedures. Thus genetic information is expressed by transcription of the **non-coding** strand of DNA, which produces an mRNA molecule that has the same sequence as the coding strand of DNA (although the RNA has uracil substituted for thymine, see Fig. 2.8(*a*)). The sequence of the coding strand is usually reported when dealing with DNA sequence data, as this permits easy reference to the sequence of the RNA.

In addition to the sequence of bases that specifies the codons in a protein-coding gene, there are other important regulatory sequences associated with genes (Fig. 2.5). A site for starting transcription is required, and this encompasses a region which binds RNA polymerase, known as the **promoter** (P), and a specific start point for transcription (T_C). A stop site for transcription (t_C) is also required. From T_C start to t_C stop is sometimes called the **transcriptional unit**, i.e. the DNA region that is copied into RNA. Within this transcriptional unit there may be regulatory sites for translation, namely a start site (T_L) and a stop signal (t_L). Other sequences involved in the control of gene expression may be present either upstream or downstream from the gene itself.

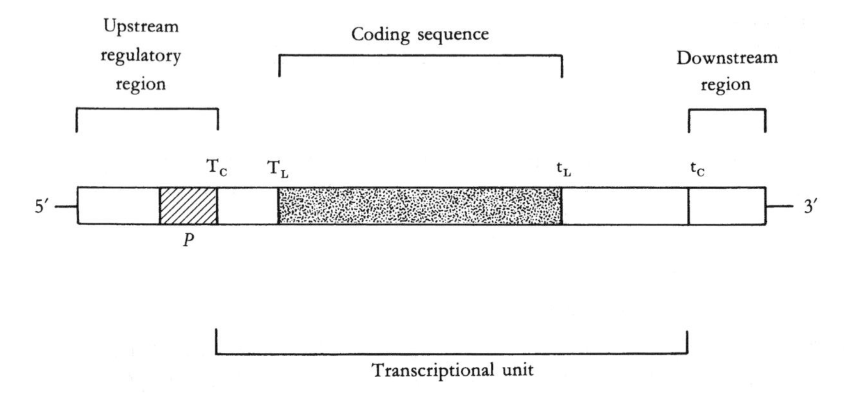

Fig. 2.5. Gene organisation. The transcriptional unit produces the RNA molecule, and is defined by the transcription start site (T_C) and stop site (t_C). Within the transcriptional unit lies the coding sequence, from the translation start site (T_L) to the stop site (t_L). The upstream regulatory region may have controlling elements such as enhancers or operators in addition to the promoter (*P*), which is the RNA polymerase-binding site.

2.3.1 Gene structure in prokaryotes

In prokaryotic cells such as bacteria, genes are usually found grouped together in **operons**. The operon is a cluster of genes that are related (often coding for enzymes in a metabolic pathway), and which are under the control of a single promoter/regulatory region. Perhaps the best known example of this arrangement is the *lac* operon (Fig. 2.6), which codes for the enzymes responsible for lactose catabolism. Within the operon there are three genes that code for proteins (termed **structural** genes) and an upstream control region encompassing the promoter and a regulatory site called the **operator**. In this control region there is also a site which binds a complex of cyclic AMP and CRP (cyclic AMP receptor protein), which is important in positive regulation (stimulation) of transcription. Lying outside the operon itself is the repressor gene, which codes for a protein (the Lac repressor) that binds to the operator site and is responsible for negative control of the operon by blocking the binding of RNA polymerase.

The fact that structural genes in prokaryotes are often grouped together means that the transcribed mRNA may contain information for more than one protein. Such a molecule is known as a **polycistronic** mRNA. Thus much of the genetic information in bacteria is expressed *via* polycistronic mRNAs whose synthesis is regulated in accordance with the needs of the cell at any given time. This system is flexible and efficient, and enables the cell to adapt quickly to changing environmental conditions.

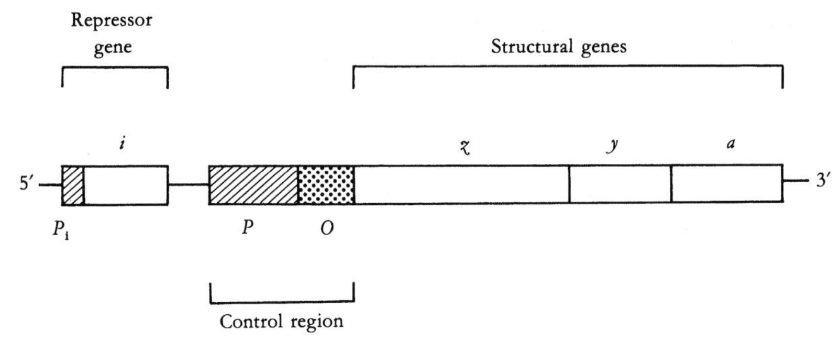

Fig. 2.6. The *lac* operon. The structural genes *lacZ*, *lacY* and *lacA* (noted as *z*, *y* and *a*) encode β-galactosidase, galactoside permease and a transacetylase, respectively. The cluster is controlled by a promoter (*P*) and an operator region (*O*). The operator is the binding site for the repressor protein, encoded by the *lacI* gene (*i*). The repressor gene lies outside the operon itself and is controlled by its own promoter, P_i.

2.3.2 Gene structure in eukaryotes

A major defining feature of eukaryotic cells is the presence of a membrane-bound nucleus, within which the DNA is stored in the form of chromosomes. Transcription therefore occurs within the nucleus, and is separated from the site of translation, which is in the cytoplasm. The picture is complicated further by the presence of genetic information in mitochondria (plant and animal cells) and chloroplasts (plant cells only), which have their own separate genomes that specify many of the components required by these organelles. This compartmentalisation has important consequences for regulation, both genetic and metabolic, and thus gene structure and function in eukaryotes is more complex than in prokaryotes.

The most startling discovery concerning eukaryotic genes was made in 1977, when it became clear that eukaryotic genes contained 'extra' pieces of DNA that did not appear in the mRNA that the gene encoded. These sequences are known as **intervening sequences** or **introns**, with the sequences that will make up the mRNA being called **exons**. In many cases the number and total length of the introns exceeds that of the exons, as in the chicken ovalbumin gene, which has a total of seven introns making up more than 75% of the gene. As our knowledge has developed, it has become clear that eukaryotic genes are often extremely complex, and may be very large indeed. Some examples of human gene complexity are shown in Table 2.2. This illustrates the tremendous range of sizes for human genes, the smallest of which may be only a few hundred base-pairs in length. At the other end of the scale, the dystrophin gene is spread over 2.4 Mb of DNA

Table 2.2. *Size and structure of some human genes*

Gene	Gene size (kbp)	Number of exons	% exon
Insulin	1.4	3	33
β-globin	1.6	3	38
Serum albumin	18	14	12
Blood clotting factor VIII	186	26	3
CFTR (cystic fibrosis)	230	27	2.4
Dystrophin (muscular dystrophy)	2400	79	0.6

Note: Gene sizes are given in kilobase-pairs (kbp). The number of exons is shown, and the percentage of the gene that is represented by these exons is given in the final column.

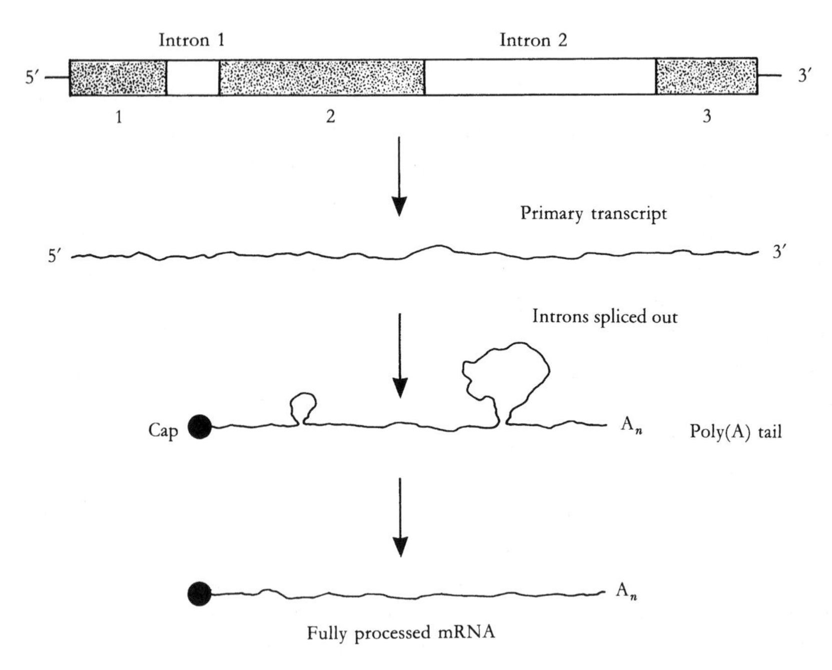

Fig. 2.7. Structure and expression of the mammalian *β*-globin gene. The gene contains two intervening sequences or introns. The expressed sequences (exons) are shaded and numbered. The primary transcript is processed by capping, polyadenylation and splicing to yield the fully functional mRNA.

on the X chromosome, with the 79 exons representing only 0.6% of this length of DNA.

The presence of introns obviously has important implications for the expression of genetic information in eukaryotes, in that the introns must be removed before the mRNA can be translated. This is carried out in the nucleus, where the introns are spliced out of the primary transcript. Further intranuclear modification includes the addition of a 'cap' at the 5′ terminus and a 'tail' of adenine residues at the 3′ terminus. These modifications are part of what is known as RNA processing, and the end product is a fully functional mRNA that is ready for export to the cytoplasm for translation. The structures of the mammalian β-globin gene and its processed mRNA are outlined in Fig. 2.7 to illustrate eukaryotic gene structure and RNA processing.

2.4 Gene expression

As shown in Fig. 2.1, the flow of genetic information is from DNA to protein. Whilst a detailed knowledge of gene expression is not required in order to understand the principles of genetic engineering, it is important to be familiar with the basic features of transcription and translation. A brief description of these processes is given here.

Transcription involves synthesis of an RNA from the DNA template provided by the non-coding strand of the transcriptional unit in question. The enzyme responsible is **RNA polymerase** (DNA-dependent RNA polymerase). In prokaryotes there is a single RNA polymerase enzyme, but in eukaryotes there are three types of RNA polymerase (I, II and III). These synthesise ribosomal, messenger and transfer/5 S ribosomal RNAs respectively. All RNA polymerases are large multisubunit proteins with relative molecular masses of around 500 000.

Transcription has several component stages, these being (i) DNA/RNA polymerase binding, (ii) chain initiation, (iii) chain elongation, and (iv) chain termination and release of the RNA. Promoter structure is important in determining the binding of RNA polymerase, but will not be dealt with here. When the RNA molecule is released, it may be immediately available for translation (as in prokaryotes) or it may be processed and exported to the cytoplasm (as in eukaryotes) before translation occurs.

Translation requires an mRNA molecule, a supply of charged tRNAs (tRNA molecules with their associated amino acid residues) and ribosomes (composed of rRNA and ribosomal proteins). The ribosomes are the sites where protein synthesis occurs; in prokaryotes, ribosomes are composed of

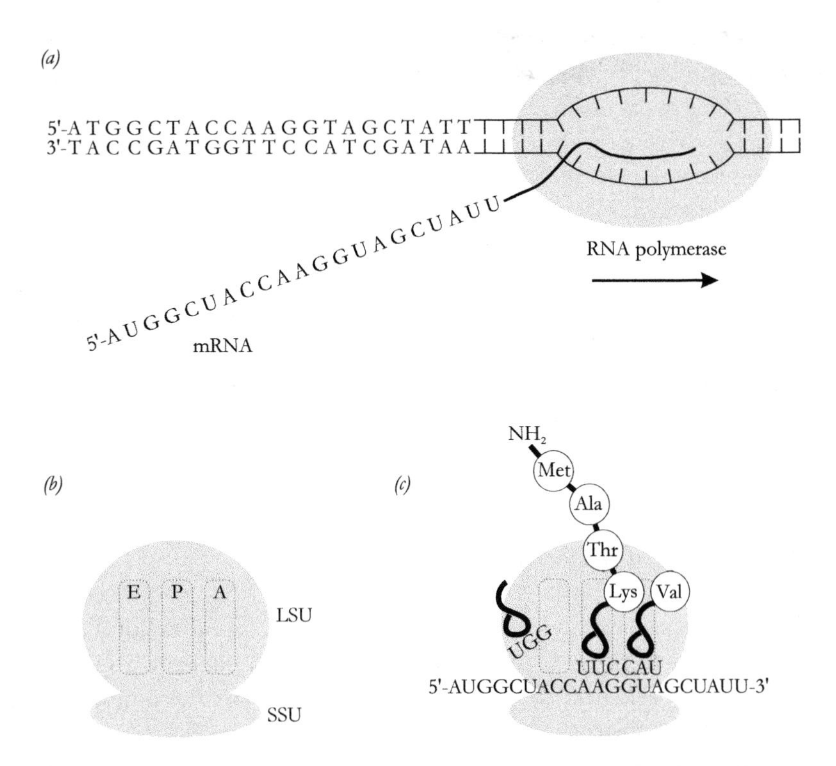

Fig. 2.8. Transcription and translation. (*a*) Transcription involves synthesis of mRNA by RNA polymerase. Part of the DNA/mRNA sequence is given. The mRNA has the same sequence as the coding strand in the DNA (the non-template strand), apart from U being substituted for T. (*b*) The ribosome is the site of translation, and is made up of the large subunit (LSU) and the small subunit (SSU), each made up of ribosomal RNA molecules and many different proteins. There are three sites within the ribosome. The A (aminoacyl) and P (peptidyl) sites are involved in insertion of the correct tRNA–amino acid complex in the growing polypeptide chain. The E (exit) site facilitates the release of the tRNA after peptide bond formation has removed its amino acid. (*c*) The mRNA is being translated. The amino acid residue is inserted into the protein in response to the codon/anticodon recognition event in the ribosome. The first amino acid residue is encoded by AUG in the mRNA (tRNA anticodon TAC), which specifies methionine (see Table 2.1 for the genetic code). The remainder of the sequence is translated in a similar way. The ribosome translates the mRNA in a 5′ 3′ direction, with the polypeptide growing from its N terminus. The residues in the polypeptide chain are joined together by peptide bonds.

three rRNAs and some 52 different ribosomal proteins. The ribosome is a complex structure that essentially acts as a 'jig' which holds the mRNA in place so that the **codons** may be matched up with the appropriate **anticodon** on the tRNA, thus ensuring that the correct amino acid is inserted into the growing polypeptide chain. The mRNA molecule is translated in a 5′ 3′ direction, corresponding to polypeptide elongation from N terminus to C terminus.

Although transcription and translation are complex processes, the essential features (with respect to information flow) may be summarised as shown in Fig. 2.8. In conjunction with the brief descriptions presented above, this should provide enough background information about gene structure and expression to enable subsequent sections of the text to be linked to these processes where necessary.

2.5 Genes and genomes

When techniques for the examination of DNA became established, gene structure was naturally one of the first areas where efforts were concentrated. However, genes do not exist in isolation, but as part of the **genome** of an organism. Over the past few years the emphasis in molecular biology has shifted slightly, and today we are much more likely to consider the genome as a whole – almost as a type of cellular organelle – rather than just a collection of genes. The human genome project (considered in Chapter 9) is a good example of the development of the field of bioinformatics, which is one of the most active research areas in modern molecular biology.

2.5.1 Genome size and complexity

The amount of DNA in the haploid genome is known as the C-value. It would seem reasonable to assume that genome size should increase with increasing complexity of organisms, reflecting the greater number of genes required to facilitate this complexity. The data shown in Table 2.3 show that, as expected, genome size does tend to increase with organismal complexity. Thus bacteria, yeast, fruit fly and human genomes fit this pattern. However, mouse, tobacco and wheat have much larger genomes than humans – this seems rather strange, as intuitively we might assume that a wheat plant is not as complex as a human being. Also, as *E. coli* has around 4000 genes, it appears that the tobacco plant genome has the capacity to encode 4 000 000 genes, and this is

Table 2.3. *Genome size in some organisms*

Organism	Genome size (Mb)
Escherichia coli (bacterium)	4.6
Saccharomyces cerevisiae (yeast)	12.1
Drosophila melanogaster (fruit fly)	150
Homo sapiens (man)	3000
Mus musculus (mouse)	3300
Nicotiana tabacum (tobacco)	4500
Triticum aestivum (wheat)	17 000

Note: Genome sizes are given in Megabase-pairs (1 Megabase $= 1 \times 10^6$ bases).

certainly not the case, even allowing for the increased size and complexity of eukaryotic genes. This anomaly is sometimes called the **C-value paradox.**

In addition to size of the genome, genome complexity also tends to increase with more complex organisation. One way of studying complexity involves examining the renaturation of DNA samples. If a DNA duplex is denatured by heating the solution until the strands separate, the complementary strands will renature on cooling (Fig. 2.9). This feature can be used to provide information about the sequence complexity of the DNA in question, since sequences that are present as multiple copies in the genome will renature faster than sequences that are present as single copies only. By performing this type of analysis, eukaryotic DNA can be shown to be composed of four different abundance classes. Firstly, some DNA will form duplex structures almost instantly, because the denatured strands have regions such as **inverted repeats** or **palindromes**, which fold back on each other to give a hairpin loop structure. This class is commonly known as **foldback DNA.** Secondly, fastest to re-anneal are **highly repetitive** sequences, which occur many times in the genome. Following these are **moderately repetitive** sequences, and finally there are the **unique** or **single copy** sequences, which rarely re-anneal under the conditions used for this type of analysis. We will consider how repetitive DNA sequence elements can be used in genome mapping and DNA profiling in Chapters 9 and 11.

2.5.2 Genome organisation

The C-value paradox and the sequence complexity of eukaryotic genomes raise questions about how genomes are organised. Viral and bacterial genomes tend to show very efficient use of DNA for encoding their genes,

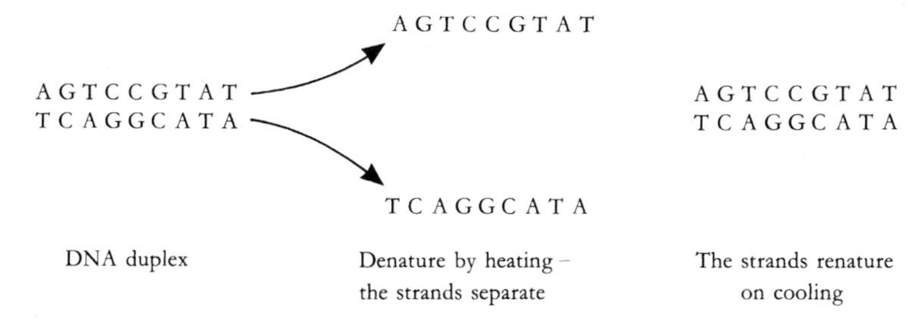

Fig. 2.9. The principle of nucleic acid hybridisation. This feature of DNA molecules is a critical part of many of the procedures involved in gene manipulation.

which is a consequence of (and explanation for) their small genome size. However, in the human genome, only about 3% of the total amount of DNA is actually coding sequence. Even when the introns and control sequences are added, the majority of the DNA has no obvious function. This is sometimes termed 'junk' DNA, although this is perhaps the wrong way to think about this apparently redundant DNA.

Estimating the number of genes in a particular organism is not an exact science, and a number of different methods may be used. When the full genome sequence is determined, this obviously makes gene identification much easier, although there are many cases where gene coding sequences are recognised, but the protein products are unknown in terms of their biological function. The human nuclear genome was thought to contain between 60000 and 80000 genes, with some estimates placing the number as high as 100000 or as low as 30000. In contrast to this imprecise estimation, the human mitochondrial genome (16 600 base-pairs in length) encodes 37 genes – reflecting the fact that smaller genomes are easier to analyse in detail.

Many genes in eukaryotes are single copy genes, and tend to be dispersed across the multiple chromosomes found in eukaryotic cell nuclei. Other genes may be part of **multigene families**, and may be grouped at a particular chromosomal location or may be dispersed. When studying gene organisation in the context of the genome itself, features such as gene density, gene size, mRNA size, intergenic distance and intron/exon sizes are important indicators. In humans the 'average' size of a coding region is around 1500 base-pairs, and the average size of a gene is 10–15 kbp. Gene density is about 1 gene per 40–45 kbp, and the intergenic distance is around 25–30 kbp. However, as we have already seen, gene structure in eukaryotes can be very complex, and thus using 'average' estimates is a little misleading. What is clear is that genomes are likely to yield much new information about gene structure and function as genome-sequencing projects generate more data.

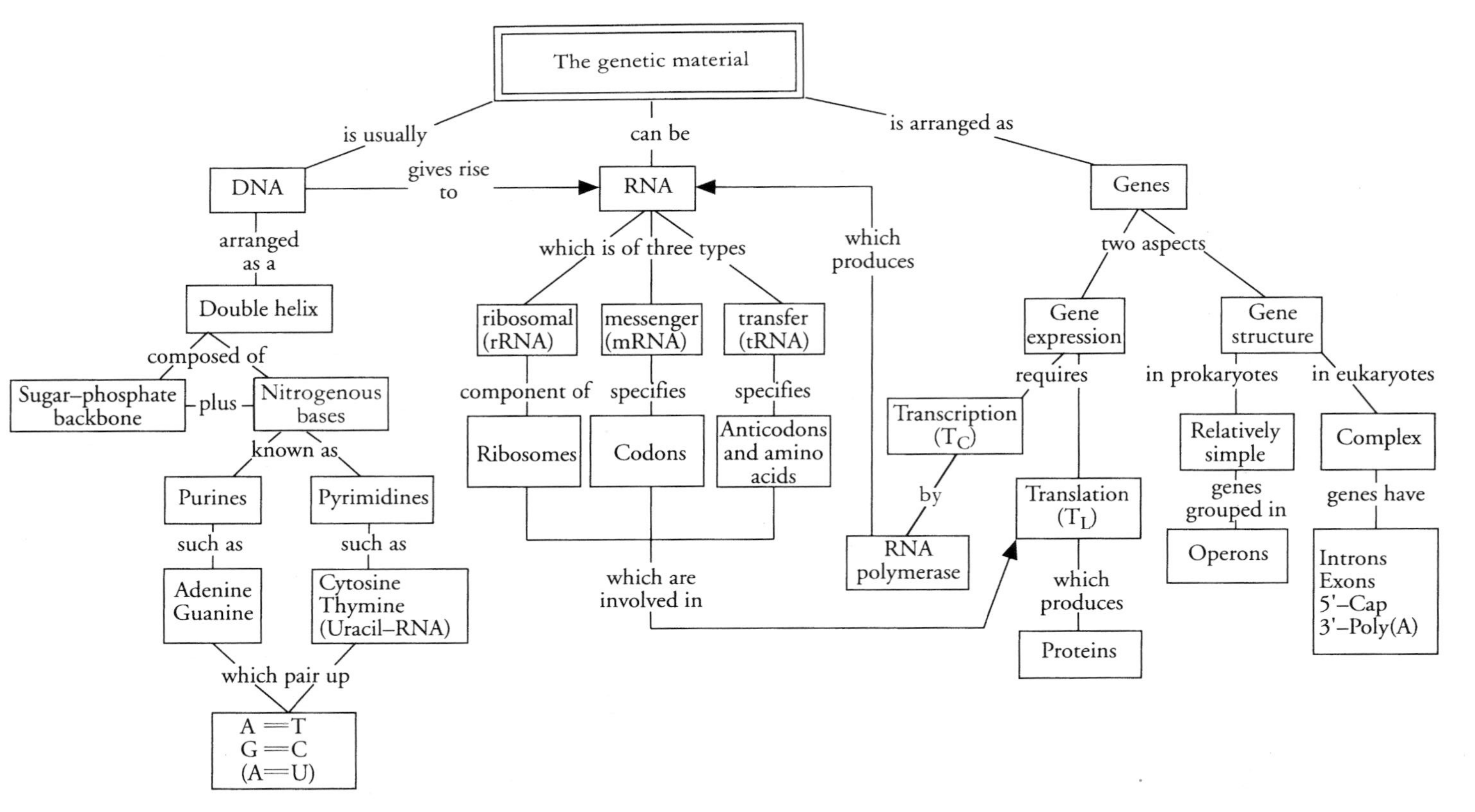

Concept map 2

3

Working with nucleic acids

Before examining some of the specific techniques used in gene manipulation, it is useful to consider the basic methods required for handling, quantifying and analysing nucleic acid molecules. It is often difficult to make the link between theoretical and practical aspects of a subject, and an appreciation of the methods used in routine work with nucleic acids may be of help when the more detailed techniques of gene cloning and analysis are described.

3.1 Isolation of DNA and RNA

Every gene manipulation experiment requires a source of nucleic acid, in the form of either DNA or RNA. It is therefore important that reliable methods are available for isolating these components from cells. There are three basic requirements: (i) opening the cells in the sample to expose the nucleic acids for further processing, (ii) separation of the nucleic acids from other cell components, and (iii) recovery of the nucleic acid in purified form. A variety of techniques may be used, ranging from simple procedures with few steps, up to more complex purifications involving several different stages. These days, most biological supply companies sell kits that enable purification of nucleic acids from a range of sources.

The first step in any isolation protocol is disruption of the starting material, which may be viral, bacterial, plant or animal. The method used to open cells should be as gentle as possible, preferably utilising enzymatic degradation of cell wall material (if present) and detergent lysis of cell membranes. If more vigorous methods of cell disruption are required (as is the case with some

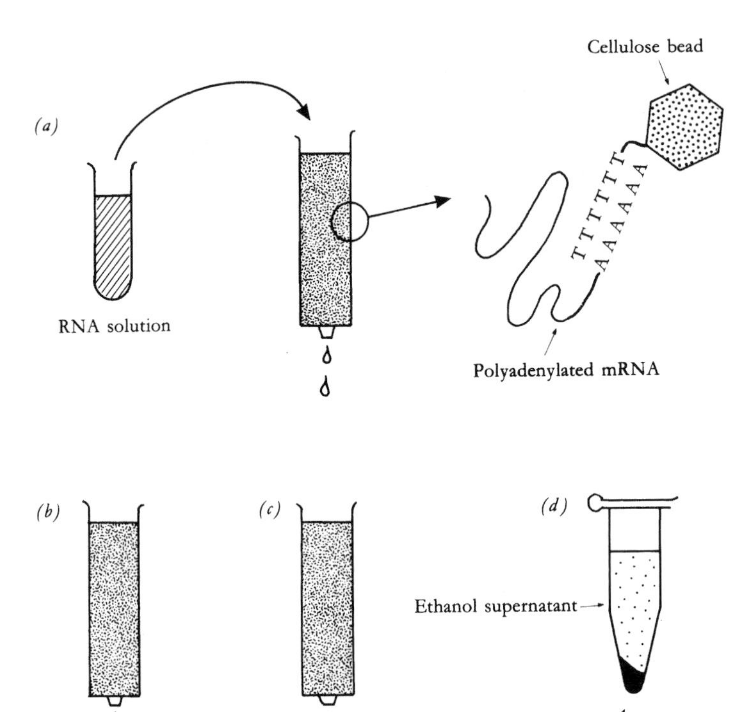

Fig. 3.1. Preparation of mRNA by affinity chromatography using oligo(dT)-cellulose. (*a*) Total RNA in solution is passed through the column in a high-salt buffer, and the oligo(dT) tracts bind the poly(A) tails of the mRNA. (*b*) Residual RNA is washed away with high-salt buffer, and (*c*) the mRNA is eluted by washing with a low-salt buffer. The mRNA is then precipitated under ethanol and collected by centrifugation (*d*).

types of plant cell material), there is the danger of shearing large DNA molecules, and this can hamper the production of representative recombinant molecules during subsequent processing.

Following cell disruption, most methods involve a deproteinisation stage. This is often achieved by one or more extractions using phenol or phenol/chloroform mixtures. On the formation of an emulsion and subsequent centrifugation to separate the phases, protein molecules partition into the phenol phase and accumulate at the interface. The nucleic acids remain mostly in the upper aqueous phase, and may be precipitated from solution using isopropanol or ethanol (see Section 3.2).

If a DNA preparation is required, the enzyme **ribonuclease** (RNase) can be used to digest the RNA in the preparation. If mRNA is needed for cDNA synthesis, a further purification can be performed by using oligo(dT)-cellulose to bind the poly(A) tails of eukaryotic mRNAs (Fig. 3.1). This gives substantial enrichment for mRNA and enables most contaminating DNA, rRNA and tRNA to be removed.

The technique of gradient centrifugation is often used to prepare DNA, particularly plasmid DNA (pDNA). In this technique a caesium chloride solution containing the DNA preparation is spun at high speed in an ultracentrifuge. Over a long period (up to 48 h in some cases) a density gradient is formed and the pDNA forms a band at one position in the centrifuge tube. The band may be taken off and the CsCl removed by dialysis to give a pure preparation of pDNA. As an alternative to gradient centrifugation, size exclusion chromatography (gel filtration) or similar techniques may be used.

3.2 Handling and quantification of nucleic acids

It is often necessary to use very small amounts of nucleic acid (typically **micro-**, **nano-** or **picograms**) during a cloning experiment. It is obviously impossible to handle these amounts directly, so most of the measurements that are done involve the use of aqueous solutions of DNA and RNA. The concentration of a solution of nucleic acid can be determined by measuring the absorbance at 260 nanometres, using a spectrophotometer. An A_{260} of 1.0 is equivalent to a concentration of 50 μg ml^{-1} for double-stranded DNA, or 40 μg ml^{-1} for single-stranded DNA or RNA. If the A_{280} is also determined, the A_{260}/A_{280} ratio indicates if there are contaminants present, such as residual phenol or protein. The A_{260}/A_{280} ratio should be around 1.8 for pure DNA and 2.0 for pure RNA preparations.

In addition to spectrophotometric methods, the concentration of DNA may be estimated by monitoring the fluorescence of bound **ethidium bromide**. This dye binds between the DNA bases (intercalates) and fluoresces orange when illuminated with ultraviolet (UV) light. By comparing the fluorescence of the sample with that of a series of standards, an estimate of the concentration may be obtained. This method can detect as little as 1–5 ng of DNA, and may be used when UV-absorbing contaminants make spectrophotometric measurements impossible. Having determined the concentration of a solution of nucleic acid, any amount (in theory) may be dispensed by taking the appropriate volume of solution. In this way nanogram or picogram amounts may be dispensed with reasonable accuracy.

Precipitation of nucleic acids is an essential technique that is used in a variety of applications. The two most commonly used precipitants are isopropanol and ethanol, ethanol being the preferred choice for most applications. When added to a DNA solution in a ratio, by volume, of 2:1 in the presence of 0.2 M salt, ethanol causes the nucleic acids to come out of solution. Although it used to be thought that low temperatures (−20 °C or −70 °C) were necessary, this is not an absolute requirement, and 0 °C appears to be adequate. After precipitation the nucleic acid can be recovered by centrifugation, which causes a pellet of nucleic acid material to form at the bottom of the tube. The pellet can be dried and the nucleic acid resuspended in the buffer appropriate to the next stage of the experiment.

3.3 Radiolabelling of nucleic acids

A major problem encountered in many cloning procedures is that of keeping track of the small amounts of nucleic acid involved. This problem is magnified at each stage of the process, because losses mean that the amount of material usually diminishes after each step. One way of tracing the material is to label the nucleic acid with a radioactive molecule (usually a deoxynucleoside triphosphate (dNTP), labelled with ^{3}H or ^{32}P), so that portions of each reaction may be counted in a scintillation counter to determine the amount of nucleic acid present.

A second application of radiolabelling is in the production of highly radioactive nucleic acid molecules for use in hybridisation experiments. Such molecules are known as radioactive **probes**, and have a variety of uses (see Sections 3.4 and 8.2). The difference between labelling for tracing purposes and labelling for probes is largely one of **specific activity**, i.e. the measure of how radioactive the molecule is. For tracing purposes, a low specific activity will suffice but for probes, a high specific activity is necessary. In probe preparation the radioactive label is usually the high-energy β-emitter ^{32}P. Some common methods of labelling nucleic acid molecules are described below.

3.3.1 End labelling

In this technique the enzyme **polynucleotide kinase** is used to transfer the terminal phosphate group of ATP onto 5′-hydroxyl termini of nucleic acid molecules. If the ATP donor is radioactively labelled, this produces a labelled

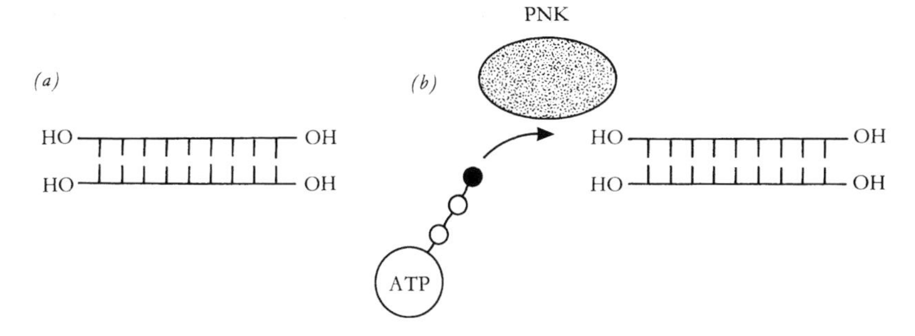

Fig. 3.2. End labelling DNA using polynucleotide kinase (PNK). (*a*) DNA is dephosphorylated using phosphatase, to generate 5′-OH groups. (*b*) The terminal phosphate of [γ-^{32}P]ATP (solid circle) is then transferred to the 5′ terminus by PNK. The reaction can also occur as an exchange reaction with 5′-phosphate termini.

nucleic acid of relatively low specific activity, as only the termini of each molecule become radioactive (Fig. 3.2).

3.3.2 Nick translation

This method relies on the ability of the enzyme **DNA polymerase I** (see Section 4.2.2) to translate (move along the DNA) a nick created in the phosphodiester backbone of the DNA double helix. Nicks may occur naturally, or may be caused by a low concentration of the nuclease **DNase I** in the reaction mixture. DNA polymerase I catalyses a strand replacement reaction which incorporates new dNTPs into the DNA chain. If one of the dNTPs supplied is radioactive, the result is a highly labelled DNA molecule (Fig. 3.3).

3.3.3 Labelling by primer extension

This term refers to a technique which uses random **oligonucleotides** (usually hexadeoxyribonucleotide molecules – sequences of six deoxynucleotides) to prime synthesis of a DNA strand by DNA polymerase. The DNA to be labelled is denatured by heating, and the oligonucleotide primers annealed to the single stranded DNAs. The **Klenow fragment** of DNA polymerase (see Section 4.2.2) can then synthesise a copy of the template, primed from the 3′-hydroxyl group of the oligonucleotide. If a labelled dNTP is incorporated, DNA of very high specific activity is produced (Fig. 3.4).

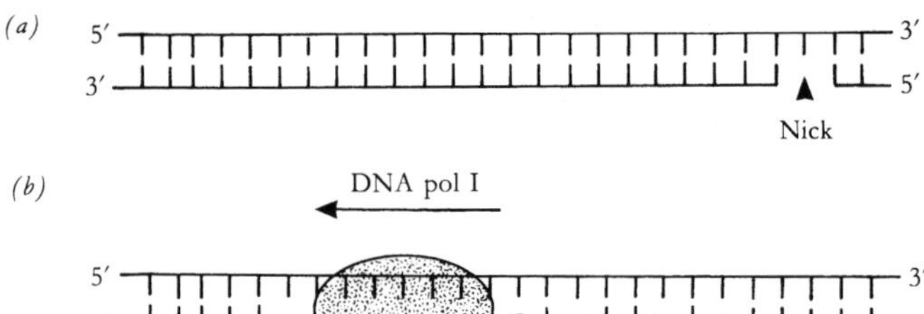

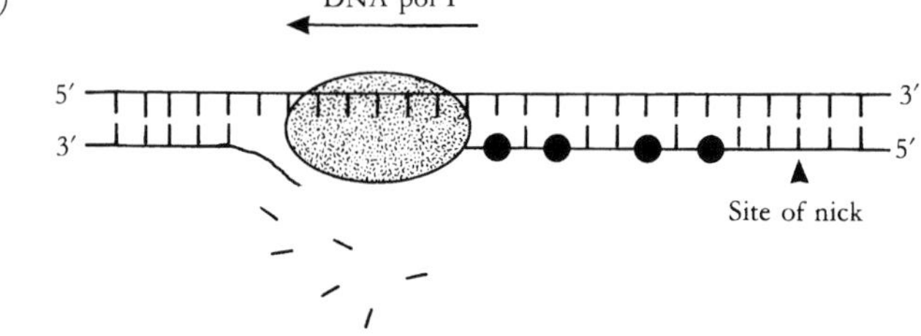

Fig. 3.3. Labelling DNA by nick translation. (*a*) A single-strand nick is introduced into the phosphodiester backbone of a DNA fragment using DNase I. (*b*) DNA polymerase I then synthesises a copy of the template strand, degrading the non-template strand with its 5′ 3′ exonuclease activity. If [α-^{32}P]dNTP is supplied this will be incorporated into the newly synthesised strand (filled circles).

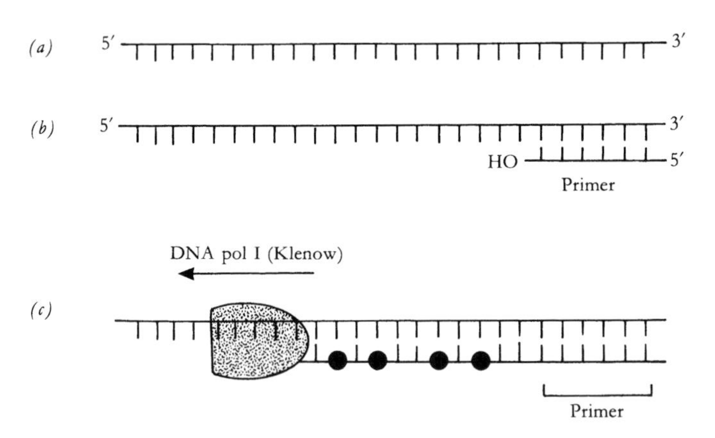

Fig. 3.4. Labelling DNA by primer extension (oligolabelling). (*a*) DNA is denatured to give single-stranded molecules. (*b*) An oligonucleotide primer is then added to give a short double-stranded region with a free 3′-OH group. (*c*) The Klenow fragment of DNA polymerase I can then synthesise a copy of the template strand from the primer, incorporating [α-^{32}P]dNTP (filled circles) to produce a labelled molecule with a very high specific activity.

In a radiolabelling reaction it is often desirable to separate the labelled DNA from the unincorporated nucleotides present in the reaction mixture. A simple way of doing this is to carry out a small-scale gel filtration step using a suitable medium. The whole process can be carried out in a Pasteur pipette, with the labelled DNA coming off the column first, followed by the free nucleotides. Fractions can be collected and monitored for radioactivity, and the data used to calculate total activity of the DNA, specific activity, and percentage incorporation of the isotope.

Radioactive labelling techniques are still used extensively in gene manipulation experiments, but pose some problems. There is the question of safety, as some high-energy radioisotopes are potentially hazardous. Thus the operator must be aware of the risks, and take appropriate precautions when using radioactive isotopes. There are also strict requirements for the disposal of radioactive waste materials. Thus there is growing use made of detection methods that use non-radioactive methods – examples include the use of fluorescent labels in DNA sequencing. We will examine some of these alternative methods later in this book.

3.4 Nucleic acid hybridisation

In addition to providing information about sequence complexity (as discussed in Section 2.5.1), nucleic acid hybridisation can be used as an extremely sensitive detection method, capable of picking out specific DNA sequences from complex mixtures. Usually a single pure sequence is labelled with ^{32}P and used as a probe. The probe is denatured before use so that the strands are free to base-pair with their complements. The DNA to be probed is also denatured, and is usually fixed to a supporting membrane made from nitrocellulose or nylon. Hybridisation is carried out in a sealed plastic bag or tube at 65–68 °C for several hours to allow the duplexes to form. The excess probe is then washed off and the degree of hybridisation can be monitored by counting the sample in a scintillation spectrometer or preparing an **autoradiogram**, where the sample is exposed to X-ray film. Some of the applications of nucleic acid hybridisation as a method for identifying cloned DNA fragments will be discussed in Chapter 8.

3.5 Gel electrophoresis

The technique of gel electrophoresis is vital to the genetic engineer, as it represents the main way by which nucleic acid fragments may be visualised

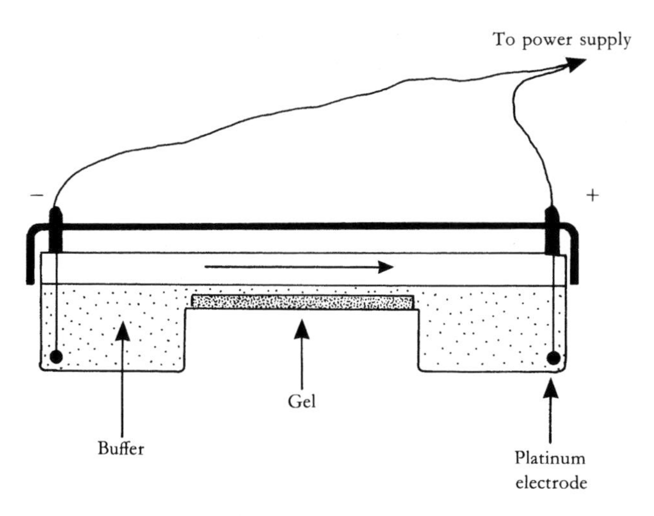

Fig. 3.5. A typical system used for agarose gel electrophoresis. The gel is just covered with buffer, therefore the technique is sometimes called *submerged agarose gel electrophoresis* (SAGE). Nucleic acid samples placed in the gel will migrate towards the positive electrode as indicated by the horizontal arrow.

directly. The method relies on the fact that nucleic acids are **polyanionic** at neutral pH, i.e. they carry multiple negative charges due to the phosphate groups on the phosphodiester backbone of the nucleic acid strands. This means that the molecules will migrate towards the positive electrode when placed in an electric field. As the negative charges are distributed evenly along the DNA molecule, the charge/mass ratio is constant, thus mobility depends on fragment length. The technique is carried out using a gel matrix, which separates the nucleic acid molecules according to size. A typical nucleic acid electrophoresis setup is shown in Fig. 3.5.

The type of matrix used for electrophoresis has important consequences for the degree of separation achieved, which is dependent on the porosity of the matrix. Two gel types are commonly used, these being **agarose** and **polyacrylamide**. Agarose is extracted from seaweed, and can be purchased as a dry powder which is melted in buffer at an appropriate concentration, normally in the range 0.3–2.0% (w/v). On cooling, the agarose sets to form the gel. Agarose gels are usually run in the apparatus shown in Fig. 3.5. Polyacrylamide gels are sometimes used to separate small nucleic acid molecules in applications such as DNA sequencing (see Section 3.6), as the pore size is smaller than that achieved with agarose. The useful separation ranges of agarose and polyacrylamide gels are shown in Table 3.1.

Electrophoresis is carried out by placing the nucleic acid samples in the gel

Table 3.1. *Separation characteristics for agarose and polyacrylamide gels*

Gel type	Separation range (base-pairs)
0.3% agarose	50000 to 1000
0.7% agarose	20000 to 300
1.4% agarose	6000 to 300
4% acrylamide	1000 to 100
10% acrylamide	500 to 25
20% acrylamide	50 to 1

Source: From Schleif and Wensink (1981). *Practical Methods in Molecular Biology*, Springer-Verlag, New York. Reproduced with permission.

and applying a potential difference across it. This is maintained until a marker dye (usually bromophenol blue, added to the sample prior to loading) reaches the end of the gel. The nucleic acids in the gel are usually visualised by staining with the intercalating dye ethidium bromide and examining under UV light. Nucleic acids show up as orange bands, which can be photographed to provide a record (Fig. 3.6). The data can be used to estimate the sizes of unknown fragments by construction of a calibration curve using standards of known size, as migration is inversely proportional to the $\log_{10}$ of the number of base-pairs. This is particularly useful in the technique of **restriction mapping** (see Section 4.1.3).

In addition to its use in the analysis of nucleic acids, polyacrylamide gel electrophoresis (PAGE) is used extensively for the analysis of proteins. The methodology is different from that used for nucleic acids, but the basic principles are similar. One common technique is SDS–PAGE, in which the detergent SDS (sodium dodecyl sulphate) is used to denature multisubunit proteins and cover the protein molecules with negative charges. In this way the inherent charge of the protein is masked, and the charge/mass ratio becomes constant. Thus proteins can be separated according to their size in a similar way to DNA molecules.

3.6 DNA sequencing

The ability to determine the sequence of bases in DNA is a central part of modern molecular biology, and provides what might be considered as the ultimate structural information. Rapid methods for sequence analysis were developed in the late 1970s, and the technique is now used in laboratories worldwide.

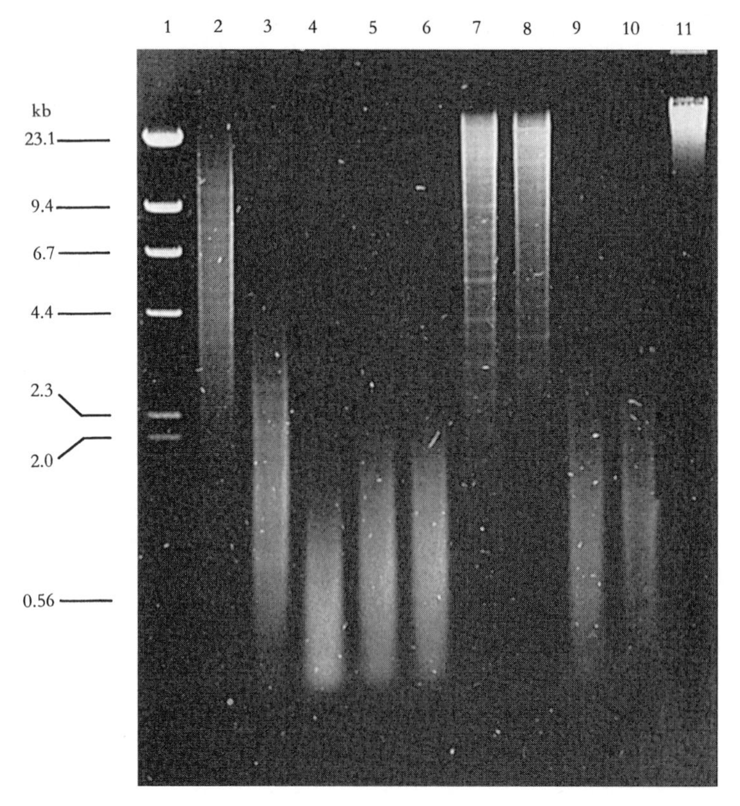

Fig. 3.6. Black-and-white photograph of an agarose gel, stained with ethidium bromide, under UV irradiation. The DNA samples show up as orange smears or as orange bands on a purple background. Individual bands (lane 1) indicate discrete fragments of DNA – in this case, the fragments are of phage λ DNA cut with the restriction enzyme *Hind*III. The sizes of the fragments (in kb) are indicated. The remaining lanes contain samples of DNA from an alga, cut with various restriction enzymes. Because of the heterogeneous nature of these samples, the fragments merge into one another and show up as a smear on the gel. Samples that have migrated farthest (lanes 3,4,5,9 and 10) are made up of smaller fragments than those that have remained near the top of the gel (lanes 2,7,8 and 11). Photograph courtesy of Dr N. Urwin.

There are two main methods for sequencing DNA. In one method, developed by Allan Maxam and Walter Gilbert, chemicals are used to cleave the DNA at certain positions, generating a set of fragments that differ by one nucleotide. The same result is achieved in a different way in the second method, developed by Fred Sanger and Alan Coulson, which involves enzymatic synthesis of DNA strands that terminate in a modified nucleotide.

Analysis of fragments is similar for both methods and involves gel electrophoresis and autoradiography (assuming that a radioactive label has been used). The enzymatic method has now largely replaced the chemical method as the technique of choice, although there are some situations where chemical sequencing can provide data more easily than the enzymatic method.

As already mentioned, fluorographic detection methods can be used in place of radioactive isotopes. This is particularly important in DNA sequencing, as it speeds up the process and enables the technique to be automated. We will look at this in more detail in Chapter 9 when considering genome sequencing.

3.6.1 Maxam–Gilbert (chemical) sequencing

A defined fragment of DNA is required as the starting material. This need not be cloned in a plasmid vector, so the technique is applicable to any DNA fragment. The DNA is radiolabelled with ^{32}P at the 5′ ends of each strand, and the strands denatured, separated and purified to give a population of labelled strands for the sequencing reactions (Fig. 3.7). The next step is a chemical modification of the bases in the DNA strand. This is done in a series of four or five reactions with different specificities, and the reaction conditions are chosen so that, on average, only one modification will be introduced into each copy of the DNA molecule. The modified bases are then removed from their sugar groups and the strands cleaved at these positions using the chemical piperidine. The theory is that, given the large number of molecules and the different reactions, this process will produce a set of fragments which terminate at different bases and differ in length by one nucleotide. This is known as a set of **nested** fragments.

3.6.2 Sanger–Coulson (dideoxy or enzymatic) sequencing

Although the end result is similar to that attained by the chemical method, the Sanger–Coulson procedure is totally different from that of Maxam and Gilbert. In this case a copy of the DNA to be sequenced is made by the Klenow fragment of DNA polymerase (see Section 4.2.2). The template for this reaction is single-stranded DNA, and a primer must be used to provide the 3′ terminus for DNA polymerase to begin synthesising the copy (Fig. 3.8). The production of nested fragments is achieved by the incorporation of a

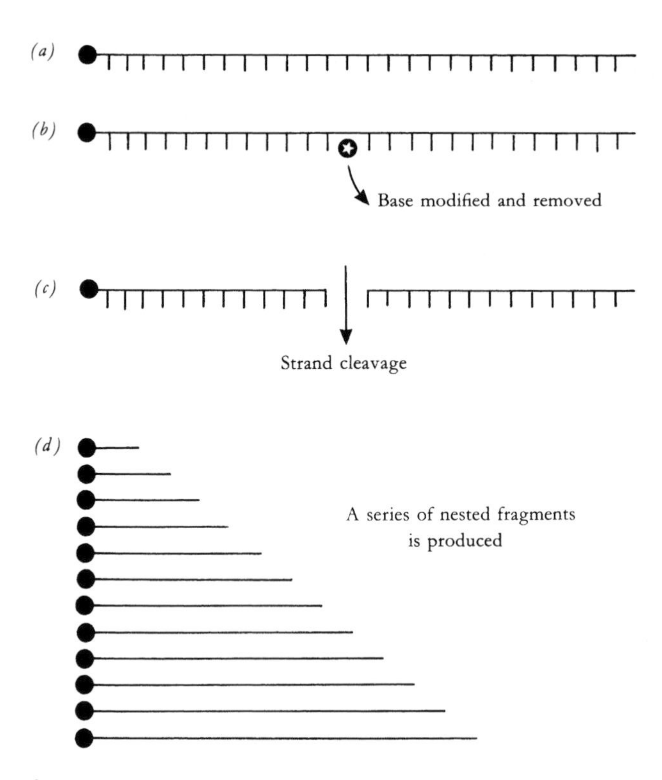

Fig. 3.7. DNA sequencing using the chemical (Maxam–Gilbert) method. (*a*) Radiolabelled single-stranded DNA is produced. (*b*) The bases in the DNA are chemically modified and removed, with, on average, one base being affected per molecule. (c) The phosphodiester backbone is then cleaved using piperidine. (*d*) The process produces a set of fragments differing in length by one nucleotide, labelled at their 5′ termini.

modified dNTP in each reaction. These dNTPs lack a hydroxyl group at the 3′ position of deoxyribose, which is necessary for chain elongation to proceed. Such modified dNTPs are known as **dideoxynucleoside triphosphates** (ddNTPs). The four ddNTPs (A,G,T and C forms) are included in a series of four reactions, each of which contains the four normal dNTPs. The concentration of the dideoxy form is such that it will be incorporated into the growing DNA chain infrequently. Each reaction therefore produces a series of fragments terminating at a specific nucleotide, and the four reactions together provide a set of nested fragments. The DNA chain is labelled by including a radioactive dNTP in the reaction mixture. This is usually [α-^{35}S]dATP, which enables more sequence to be read from a single gel than the ^{32}P-labelled dNTPs that were used previously.

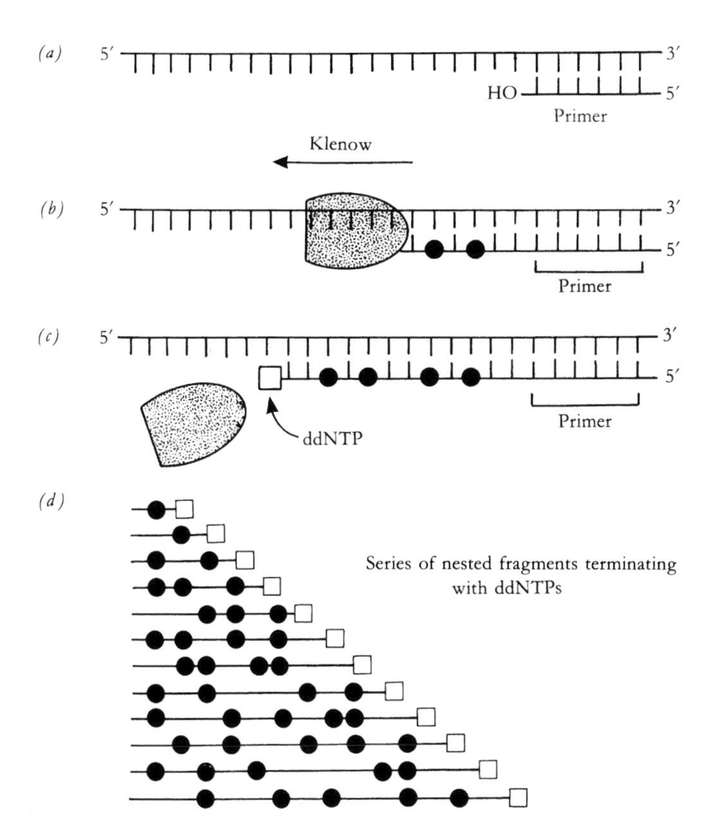

Fig. 3.8. DNA sequencing using the dideoxy chain termination (Sanger–Coulson) method. (*a*) A primer is annealed to a single-stranded template and (*b*) the Klenow fragment of DNA polymerase I used to synthesise a copy of the DNA. A radiolabelled dNTP (often [α-^{35}S]dNTP, filled circles) is incorporated into the DNA. (c) Chain termination occurs when a dideoxy nucleoside triphosphate (ddNTP) is incorporated. (*d*) A series of four reactions, each containing one ddNTP in addition to the four dNTPs required for chain elongation, generates a set of radiolabelled nested fragments.

The generation of fragments for dideoxy sequencing is more complicated than for chemical sequencing, and usually involves subcloning into different vectors. Many plasmid vectors are now available (see Section 5.2), and some types can be used directly for DNA sequencing experiments. Another method is to clone the DNA into a vector such as the bacteriophage M13 (see Section 5.3.3), which produces single-stranded DNA during infection. This provides a suitable substrate for the sequencing reactions.

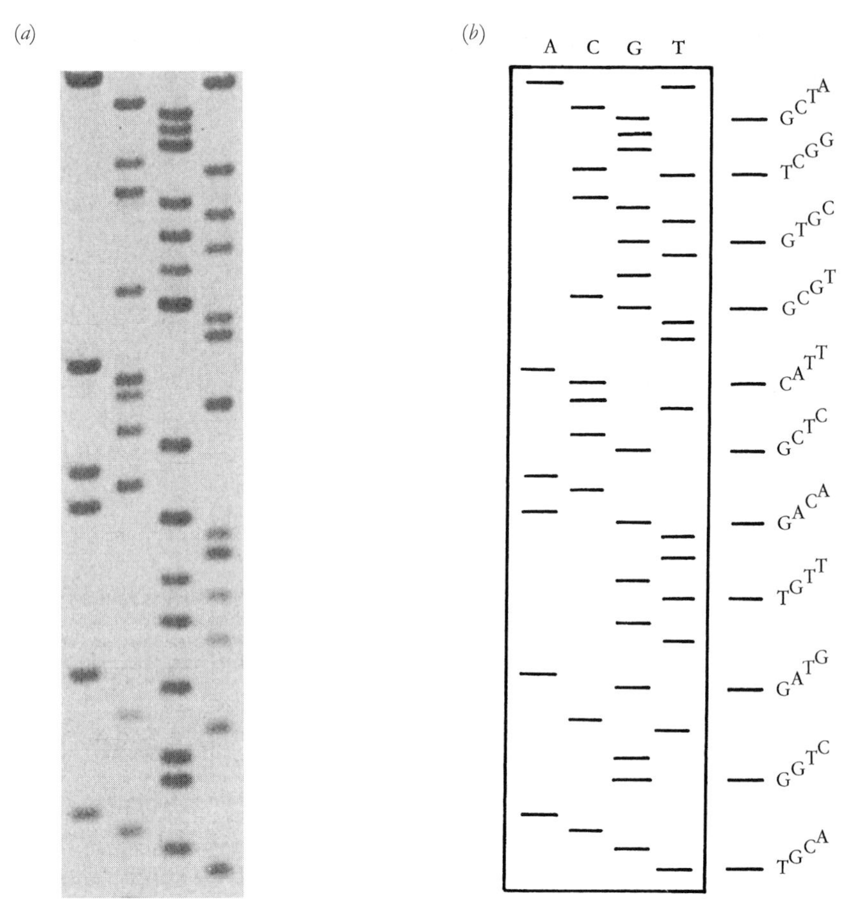

Fig. 3.9. Reading a DNA sequence. (*a*) An autoradiograph of part of a sequencing gel, and (*b*) a tracing of the autoradiograph. Each lane corresponds to a reaction containing one of the four ddNTPs. The sequence is read from the bottom of the gel, each successive fragment being one nucleotide longer than the preceding one. Photograph courtesy of Dr N. Urwin.

3.6.3 Electrophoresis and reading of sequences

Separation of the DNA fragments created in sequencing reactions is achieved by polyacrylamide gel electrophoresis. The gels usually contain 6–20% polyacrylamide and 7 M urea, which acts as a denaturant to reduce the effects of DNA secondary structure. This is important because fragments that differ in length by only one base are being separated. The gels are very thin (0.5 mm or less) and are run at high power settings, which causes them to heat up to

60–70 °C. This also helps to maintain denaturing conditions. Sometimes two lots of samples are loaded onto the same gel at different times to maximise the amount of sequence information obtained.

When the gel has been run, it is removed from the apparatus and may be dried onto a paper sheet to facilitate handling. It is then exposed to X-ray film. The emissions from the radioactive label sensitise the silver grains, which turn black when the film is developed and fixed. The result is known as an autoradiogram (Fig. 3.9(*a*)). Reading the autoradiogram is straightforward – the sequence is read from the smallest fragment upwards, as shown in Fig. 3.9(*b*). Using this method sequences of up to several hundred bases may be read from single gels. The sequence data are then compiled and studied using a computer, which can perform analyses such as translation into amino acid sequences and identification of restriction sites, regions of sequence homology and other structural motifs such as promoters and control regions.

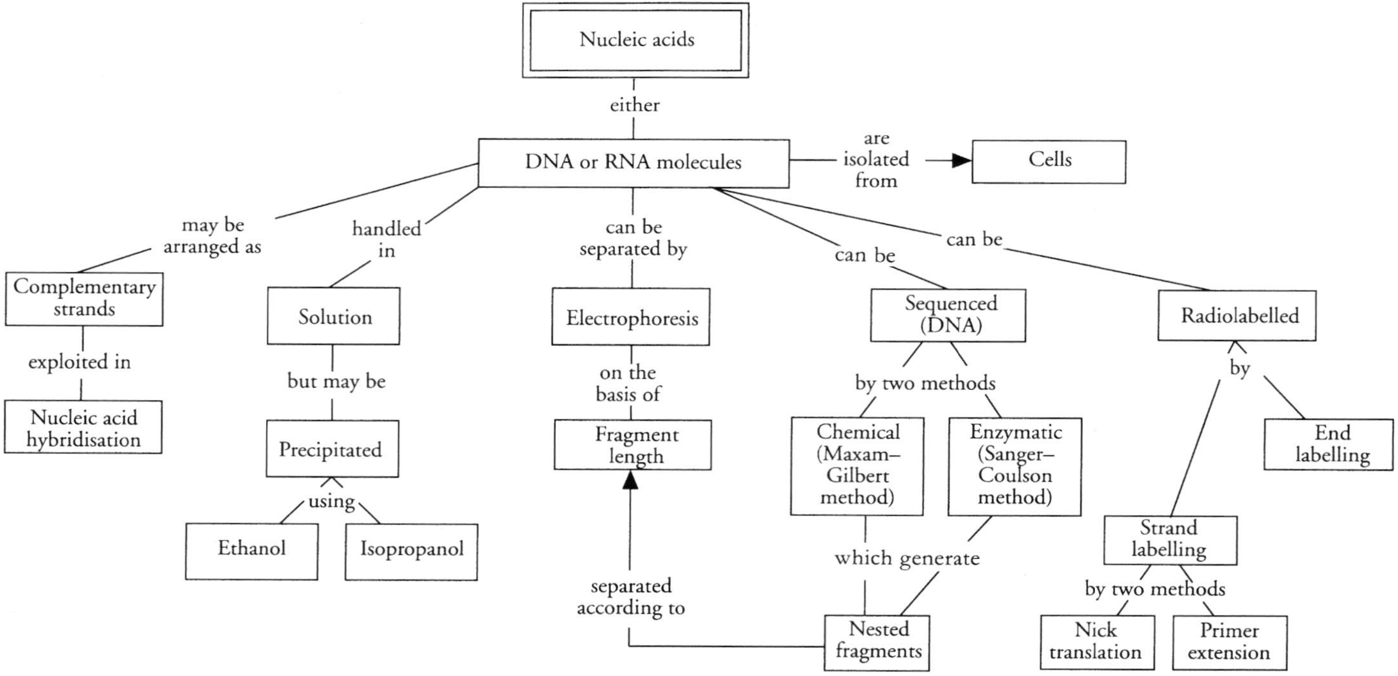

Nucleic acids
either
DNA or RNA molecules
are isolated from
Cells
may be arranged as
Complementary strands
exploited in
Nucleic acid hybridisation
handled in
Solution
but may be
Precipitated
using
Ethanol
Isopropanol
can be separated by
Electrophoresis
on the basis of
Fragment length
separated according to
can be
Sequenced (DNA)
by two methods
Chemical (Maxam–Gilbert method)
Enzymatic (Sanger–Coulson method)
which generate
Nested fragments
can be
Radiolabelled
by
End labelling
Strand labelling
by two methods
Nick translation
Primer extension

Concept map 3

4

The tools of the trade

The genetic engineer needs to be able to cut and join DNA from different sources. In addition, certain modifications may have to be carried out to the DNA during the various steps required to produce, clone and identify recombinant DNA molecules. The tools that enable these manipulations to be performed are **enzymes**, which are purified from a wide range of organisms and can be bought from various suppliers. In this chapter I examine some of the important classes of enzymes that make up the genetic engineer's toolkit.

4.1 Restriction enzymes – cutting DNA

The **restriction enzymes**, which cut DNA at defined sites, represent one of the most important groups of enzymes for the manipulation of DNA. These enzymes are found in bacterial cells, where they function as part of a protective mechanism called the **restriction–modification** system. In this system the restriction enzyme hydrolyses any exogenous DNA that appears in the cell. To prevent the enzyme acting on the host cell DNA, the modification enzyme of the system (a methylase) modifies the host DNA by methylation of particular bases in the recognition sequence, which prevents the restriction enzyme from cutting the DNA.

Restriction enzymes are of three types (I, II or III). Most of the enzymes used today are type II enzymes, which have the simplest mode of action. These enzymes are **nucleases** (see Section 4.2.1), and as they cut at an internal position in a DNA strand (as opposed to beginning degradation at one end) they are known as **endonucleases**. Thus the correct designation of such enzymes is that

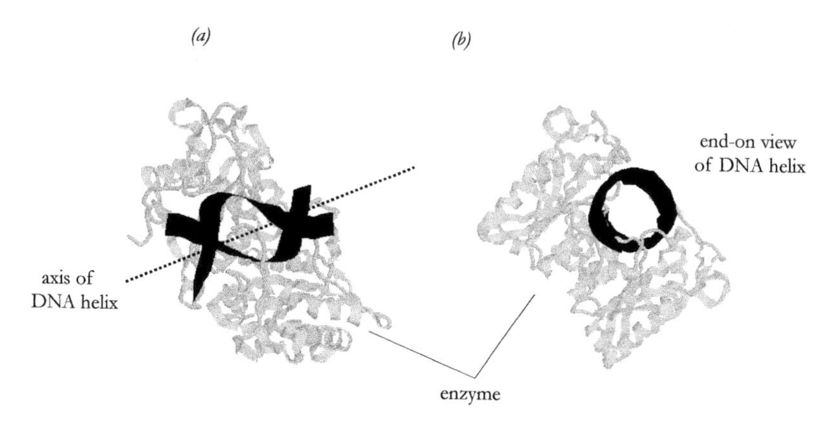

Fig. 4.1. Binding of the restriction enzyme *Bam*HI to the DNA helix. This shows how the enzyme wraps around the helix to facilitate hydrolysis of the phosphodiester linkages. (*a*) and (*b*) show different views with respect to the axis of the DNA helix. Generated using *RasMol* molecular modelling software (Roger Sayle, Public Domain). From Nicholl (2000), *Cell and Molecular Biology*, Advanced Higher Monograph Series, Learning and Teaching Scotland. Reproduced with permission.

they are type II restriction endonucleases, although they are often simply called restriction enzymes. In essence they may be thought of as molecular scissors.

4.1.1 Type II restriction endonucleases

Restriction enzyme nomenclature is based on a number of conventions. The generic and specific names of the organism in which the enzyme is found are used to provide the first part of the designation, which comprises the first letter of the generic name and the first two letters of the specific name. Thus an enzyme from a strain of *Escherichia coli* is termed *Eco*, one from *Bacillus amyloliquefaciens* is *Bam*, and so on. Further descriptors may be added, depending on the bacterial strain involved and on the presence or absence of extrachromosomal elements. Two widely used enzymes from the bacteria mentioned above are *Eco*RI and *Bam*HI. The binding of *Bam*HI to its recognition sequence is shown in Fig. 4.1.

The value of restriction endonucleases lies in their specificity. Each particular enzyme recognises a specific sequence of bases in the DNA, the most common recognition sequences being four, five or six base-pairs in length. Thus, given that there are four bases in the DNA, and assuming a random distribution of bases, the expected frequency of any particular sequence can be calculated as 4^n, where n is the length of the recognition sequence. This predicts that tetranucleotide sites will occur every 256 base-pairs, pentanucleotide

Table 4.1. *Recognition sequences and cutting sites for some restriction endonucleases*

Enzyme	Recognition sequence	Cutting sites	Ends
*Bam*HI	5′-GGATCC-3′	G↓GATCC CCTAG↑G	5′
*Eco*RI	5′-GAATTC-3′	G↓AATTC CTTAA↑G	5′
*Hae*III	5′-GGCC-3′	GG↓CC CC↑GG	Blunt
*Hpa*I	5′-GTTAAC-3′	GTT↓AAC CAA↑TTG	Blunt
*Pst*I	5′-CTGCAG-3′	CTGCA↓G G↑ACGTC	3′
*Sau*3A	5′-GATC-3′	↓GATC CTAG↑	5′
*Sma*I	5′-CCCGGG-3′	CCC↓GGG GGG↑CCC	Blunt
*Sst*I	5′-GAGCTC-3′	GAGCT↓C C↑TCGAG	3′
*Xma*I	5′-CCCGGG-3′	C↓CCGGG GGGCC↑C	5′

Note: The recognition sequences are given in single-strand form, written 5′ 3′. Cutting sites are given in double-stranded form to illustrate the type of ends produced by a particular enzyme; 5′ and 3′ refer to 5′- and 3′-protruding termini, respectively.

sites every 1024 base-pairs, and hexanucleotide sites every 4096 base-pairs. There is, as you might expect, considerable variation from these values, but generally the fragment lengths produced will lie around the calculated value. Thus an enzyme recognising a tetranucleotide sequence (sometimes called a 'four-cutter') will produce shorter DNA fragments than a six-cutter. Some of the most commonly used restriction enzymes are listed in Table 4.1, with their recognition sequences and cutting sites.

4.1.2 Use of restriction endonucleases

Restriction enzymes are very simple to use – an appropriate amount of enzyme is added to the target DNA in a buffer solution, and the reaction is incubated at 37 °C. Enzyme activity is expressed in units, with one unit being the amount of enzyme that will cleave one microgram of DNA in one hour at 37 °C. Although

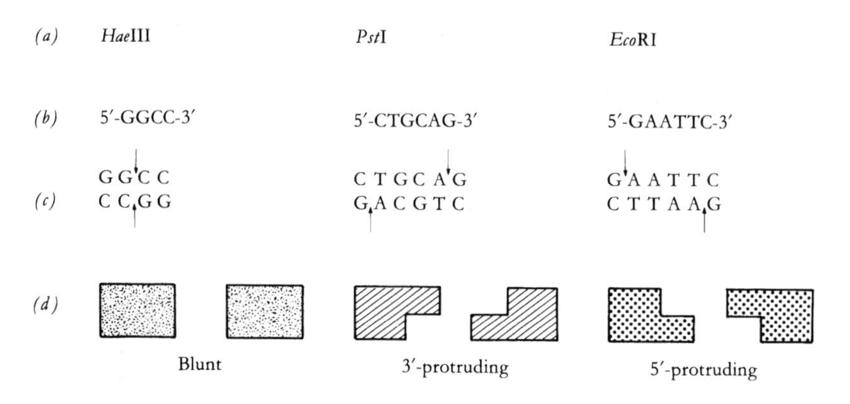

Fig. 4.2. Types of ends generated by different restriction enzymes. The enzymes are listed in (*a*), with their recognition sequences and cutting sites shown in (*b*) and (*c*), respectively. (*d*) A schematic representation of the types of ends generated.

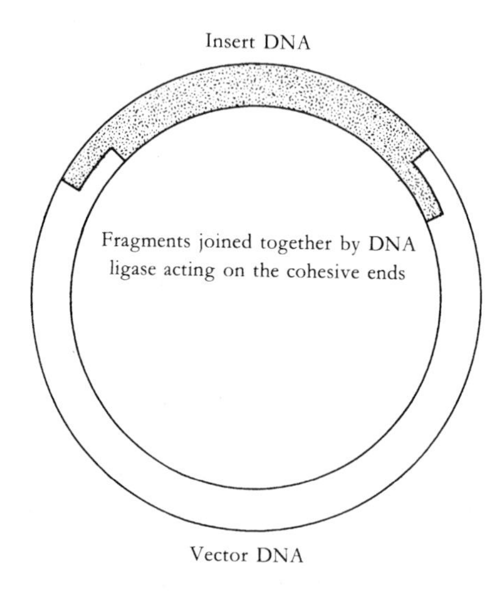

Fig. 4.3. Generation of recombinant DNA. DNA fragments from different sources can be joined together if they have cohesive ('sticky') ends, as produced by many restriction enzymes. On annealing the complementary regions, the phosphodiester backbone is sealed using DNA ligase.

most experiments require complete digestion of the target DNA, there are some cases where various combinations of enzyme concentration and incubation time may be used to achieve only partial digestion (see Section 6.3.2).

The type of DNA fragment that a particular enzyme produces depends on the recognition sequence and on the location of the cutting site within this sequence. As we have already seen, fragment length is dependent on the frequency of occurrence of the recognition sequence. The actual cutting site of the enzyme will determine the type of ends that the cut fragment has, which is important with regard to further manipulation of the DNA. Three types of fragment may be produced, these being (i) blunt or flush-ended fragments, (ii) fragments with protruding 3′ ends, and (iii) fragments with protruding 5′ ends. An example of each type is shown in Fig. 4.2.

Enzymes such as *Pst*I and *Eco*RI generate DNA fragments with cohesive or 'sticky' ends, as the protruding sequences can base-pair with complementary sequences generated by the same enzyme. Thus, by cutting two different DNA samples with the same enzyme and mixing the fragments together, recombinant DNA can be produced, as shown in Fig. 4.3. This is one of the most useful applications of restriction enzymes, and is a vital part of many manipulations in genetic engineering.

4.1.3 Restriction mapping

Most pieces of DNA will have recognition sites for various restriction enzymes, and it is often beneficial to know the relative locations of some of these sites. The technique used to obtain this information is known as **restriction mapping**. This involves cutting a DNA fragment with a selection of restriction enzymes, singly and in various combinations. The fragments produced are run on an agarose gel and their sizes determined. From the data obtained, the relative locations of the cutting sites can be worked out. A fairly simple example can be used to illustrate the technique, as outlined below.

Let us say that we wish to map the cutting sites for the restriction enzymes *Bam*HI, *Eco*RI and *Pst*I, and that the DNA fragment of interest is 15 kb in length. Various digestions are carried out, and the fragments arising from these are analysed and their sizes determined. The results obtained are shown in Table 4.2. As each of the single enzyme reactions produces two DNA fragments, we can conclude that the DNA has a single cutting site for each enzyme. The double digests enable a map to be drawn up, and the triple digest confirms this. Construction of the map is outlined in Fig. 4.4.

Table 4.2. *Digestion of a 15 kb DNA fragment with three restriction enzymes*

*Bam*HI	*Eco*RI	*Pst*I	*Bam*HI + *Eco*RI	*Bam*HI + *Pst*I	*Eco*RI + *Pst*I	*Bam*HI + *Eco*RI + *Pst*I
14	12	8	11	8	7	6
1	3	7	3	6	5	5
			1	1	3	3
						1

Note: Data shown are lengths (in kb) of fragments that are produced on digestion of a 15 kb DNA fragment with the enzymes *Bam*HI, *Eco*RI and *Pst*I. Single, double and triple digests were carried out as indicated.

4.2 DNA modifying enzymes

Restriction enzymes (described above) and DNA ligase (Section 4.3) provide the cutting and joining functions that are essential for the production of recombinant DNA molecules. Other enzymes used in genetic engineering may be loosely termed DNA **modifying** enzymes, with the term used here to include degradation, synthesis and alteration of DNA. Some of the most commonly used enzymes are described below.

4.2.1 Nucleases

Nuclease enzymes degrade nucleic acids by breaking the phosphodiester bond that holds the nucleotides together. Restriction enzymes are good examples of **endonucleases**, which cut within a DNA strand. A second group of nucleases, which degrade DNA from the termini of the molecule, are known as **exonucleases**.

Apart from restriction enzymes, there are four useful nucleases that are often used in genetic engineering. These are **Bal 31** and **exonuclease III** (exonucleases); and **deoxyribonuclease I** (DNase I) and **S_1-nuclease** (endonucleases). These enzymes differ in their precise mode of action, and provide the genetic engineer with a variety of strategies for attacking DNA. Their features are summarised in Fig. 4.5.

In addition to DNA-specific nucleases, there are **ribonucleases**, which act

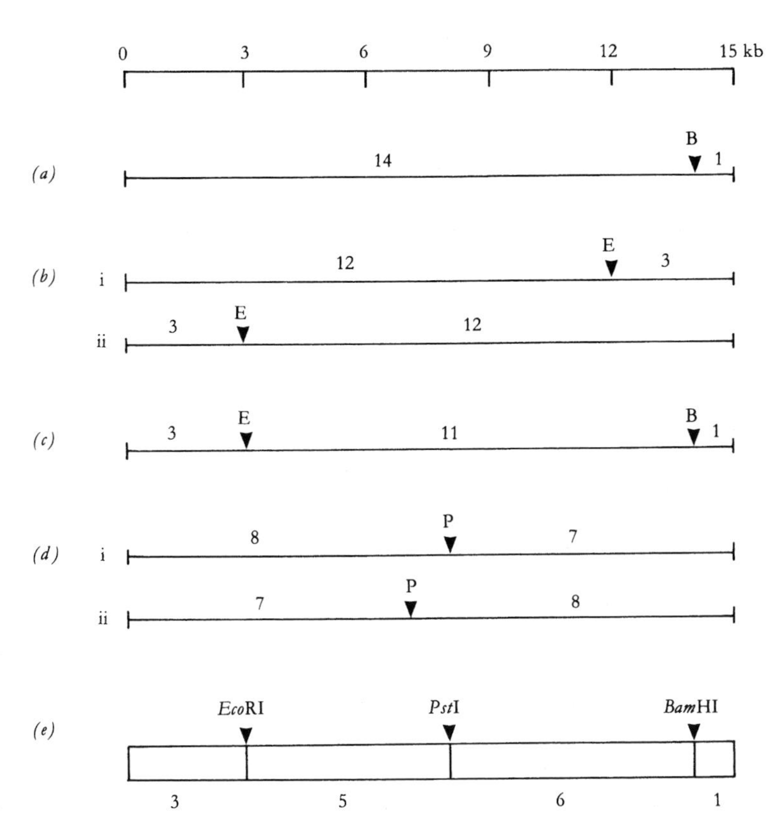

Fig. 4.4. Restriction mapping. (*a*) The 15 kb fragment yields two fragments of 14 and 1 kb when cut with *Bam*HI. (*b*) The *Eco*RI fragments of 12 and 3 kb can be orientated in two ways with respect to the *Bam*HI site. The *Bam*HI/*Eco*RI double digest gives fragments of 11, 3 and 1 kb, and therefore the relative positions of the *Bam*HI and *Eco*RI sites are as shown in (*c*). Similar reasoning with the orientation of the 8 and 7 kb *Pst*I fragments (*d*) gives the final map (*e*).

on RNA. These may be required for many of the stages in the preparation and analysis of recombinants, but as they are not directly involved in the construction of recombinant DNA molecules, they will not be described in detail.

4.2.2 Polymerases

Polymerase enzymes synthesise copies of nucleic acid molecules, and are used in many genetic engineering procedures. When describing a polymerase enzyme, the terms 'DNA-dependent' or 'RNA-dependent' may be used to

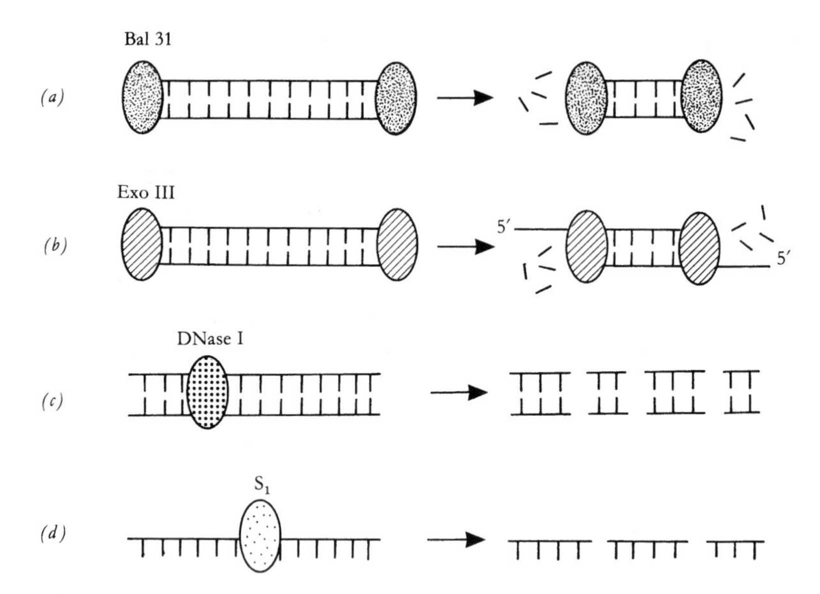

Fig. 4.5. Mode of action of various nucleases. (*a*) Nuclease Bal 31 is a complex enzyme. Its primary activity is a fast-acting 3′ exonuclease, which is coupled with a slow-acting endonuclease. When Bal 31 is present at a high concentration these activities effectively shorten DNA molecules from both termini. (*b*) Exonuclease III is a 3′ exonuclease that generates molecules with protruding 5′ termini. (*c*) DNase I cuts either single-stranded or double-stranded DNA at essentially random sites. (*d*) Nuclease S_1 is specific for single-stranded RNA or DNA. Modified from Brown (1990), *Gene Cloning*, Chapman and Hall; and Williams and Patient (1988), *Genetic Engineering*, IRL Press. Reproduced with permission.

indicate the type of nucleic acid template that the enzyme uses. Thus a DNA-dependent DNA polymerase copies DNA into DNA, an RNA-dependent DNA polymerase copies RNA into DNA, and a DNA-dependent RNA polymerase transcribes DNA into RNA. These enzymes synthesise nucleic acids by joining together nucleotides whose bases are complementary to the template strand bases. The synthesis proceeds in a 5′ 3′ direction, as each subsequent nucleotide addition requires a free 3′-OH group for the formation of the phosphodiester bond. This requirement also means that a short double-stranded region with an exposed 3′-OH (a primer) is necessary for synthesis to begin.

The enzyme **DNA polymerase I** has, in addition to its polymerase function, 5′ 3′ and 3′ 5′ exonuclease activities. The enzyme catalyses a strand replacement reaction, where the 5′ 3′ exonuclease function degrades the non-template strand as the polymerase synthesises the new copy. A major use of this enzyme is in the nick translation procedure for radiolabelling DNA (outlined in Section 3.3.2).

The 5′ 3′ exonuclease function of DNA polymerase I can be removed by cleaving the enzyme to produce what is known as the **Klenow fragment**. This retains the polymerase and 3′ 5′ exonuclease activities. The Klenow fragment is used where a single-stranded DNA molecule needs to be copied; because the 5′ 3′ exonuclease function is missing, the enzyme cannot degrade the non-template strand of dsDNA during synthesis of the new DNA. The 3′ 5′ exonuclease activity is supressed under the conditions normally used for the reaction. Major uses for the Klenow fragment include radiolabelling by primed synthesis and DNA sequencing by the dideoxy method (see Sections 3.3.3 and 3.6.2) in addition to the copying of single-stranded DNAs during the production of recombinants.

Reverse transcriptase (RTase) is an RNA-dependent DNA polymerase, and therefore produces a DNA strand from an RNA template. It has no associated exonuclease activity. The enzyme is used mainly for copying mRNA molecules in the preparation of cDNA (**complementary** or **copy** DNA) for cloning (see Section 6.2.1), although it will also act on DNA templates.

4.2.3 Enzymes that modify the ends of DNA molecules

The enzymes **alkaline phosphatase, polynucleotide kinase** and **terminal transferase** act on the termini of DNA molecules, and provide important functions that are used in a variety of ways. The phosphatase and kinase enzymes, as their names suggest, are involved in the removal or addition of phosphate groups. Bacterial alkaline phosphatase (BAP; there is also a similar enzyme, calf intestinal alkaline phosphatase, CIP) removes phosphate groups from the 5′ ends of DNA, leaving a 5′-OH group. The enzyme is used to prevent unwanted ligation of DNA molecules, which can be a problem in certain cloning procedures. It is also used prior to the addition of radioactive phosphate to the 5′ ends of DNAs by polynucleotide kinase (see Section 3.3.1).

Terminal transferase (terminal deoxynucleotidyl transferase) repeatedly adds nucleotides to any available 3′ terminus. Although it works best on protruding 3′ ends, conditions can be adjusted so that blunt-ended or 3′-recessed molecules may be utilised. The enzyme is mainly used to add homopolymer tails to DNA molecules prior to the construction of recombinants (see Section 6.2.2).

4.3 DNA ligase – joining DNA molecules

DNA ligase is an important cellular enzyme, as its function is to repair broken phosphodiester bonds that may occur at random or as a consequence of DNA replication or recombination. In genetic engineering it is used to seal discontinuities in the sugar–phosphate chains that arise when recombinant DNA is made by joining DNA molecules from different sources. It can therefore be thought of as molecular glue, which is used to stick pieces of DNA together. This function is crucial to the success of many experiments, and DNA ligase is therefore a key enzyme in genetic engineering.

The enzyme used most often in experiments is T4 DNA ligase, which is purified from *E. coli* cells infected with bacteriophage T4. Although the enzyme is most efficient when sealing gaps in fragments that are held together by cohesive ends, it will also join blunt-ended DNA molecules together under appropriate conditions. The enzyme works best at 37 °C, but is used at much lower temperatures (4–15 °C) to prevent thermal denaturation of the short base-paired regions that hold the cohesive ends of DNA molecules together.

The ability to cut, modify and join DNA molecules gives the genetic engineer the freedom to create recombinant DNA molecules. The technology involved is a test-tube technology, with no requirement for a living system. However, once a recombinant DNA fragment has been generated *in vitro*, it usually has to be amplified so that enough material is available for subsequent manipulation and analysis. Amplification usually requires a biological system, unless the **polymerase chain reaction** (**PCR**) is used (see Chapter 7 for detailed discussion of PCR). We must therefore examine the types of living systems that can be used for the propagation of recombinant DNA molecules. These systems are described in the next chapter.

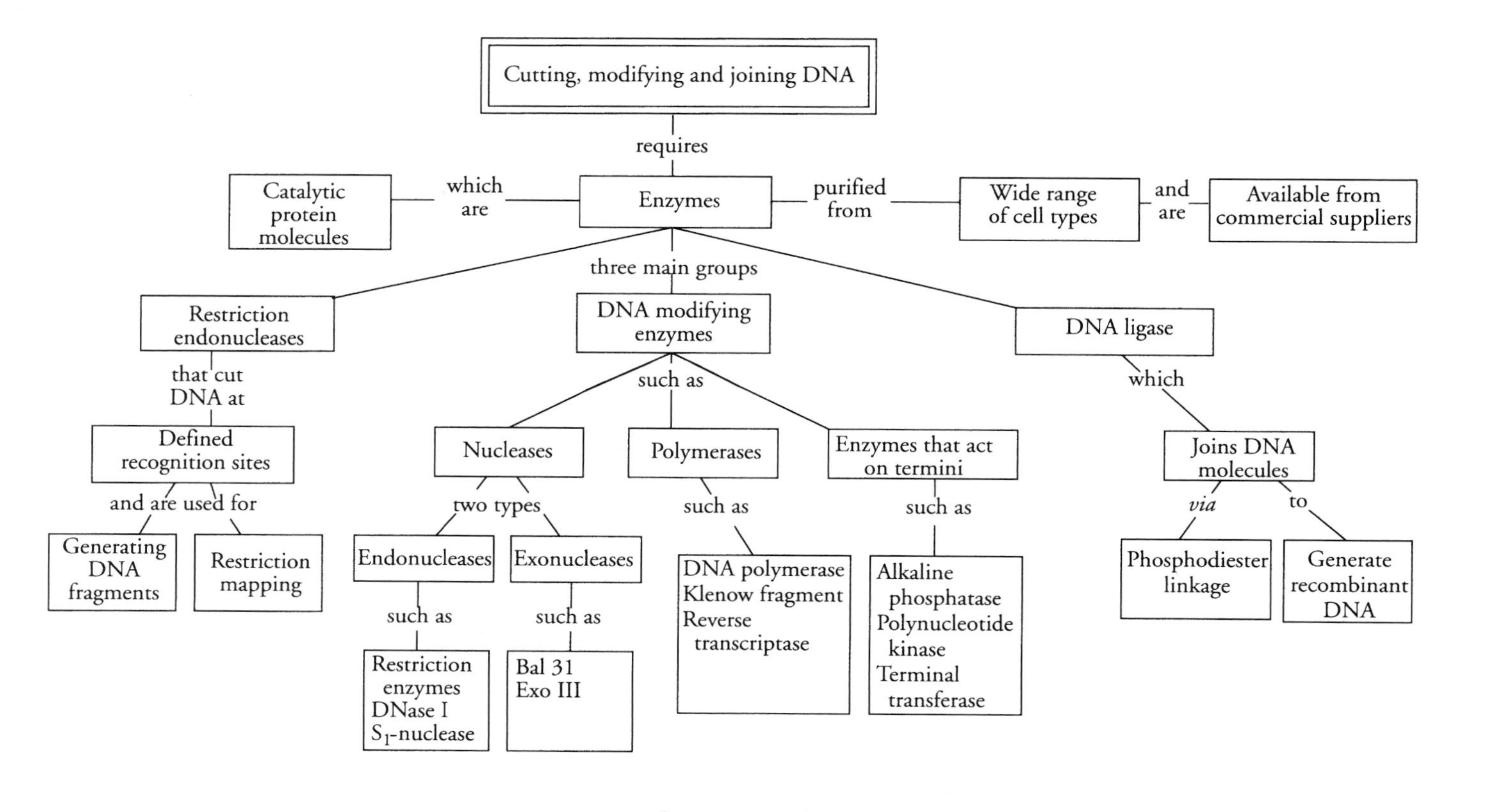

Concept map 4

Part II

The methodology of gene manipulation

5

Host cells and vectors

Once recombinant DNA molecules have been constructed *in vitro*, the desired sequence can be isolated. In some experiments hundreds of thousands of different DNA fragments may be produced, and the isolation of a particular sequence would seem to be an almost impossible task. It is a bit like looking for the proverbial needle in a haystack – with the added complication that the needle is made of the same material as the haystack! Fortunately the methods available provide a relatively simple way to isolate specific gene sequences.

Three things have to be done to isolate a gene from a collection of recombinant DNA sequences: (i) the individual recombinant molecules have to be physically separated from each other, (ii) the recombinant sequences have to be amplified to provide enough material for further analysis, and (iii) the specific fragment of interest has to be selected by some sort of sequence-dependent method. In this chapter I consider the first two of these requirements, which in essence represent the systems and techniques involved in **gene cloning**. This is an essential part of most genetic manipulation programmes. Even if the desired result is a transgenic organism, the gene to be used must first be isolated and characterised, and therefore cloning systems are required. Methods for selecting specific sequences are described in Chapter 8.

The biology of gene cloning is concerned with the selection and use of a suitable carrier molecule or **vector**, and a living system or **host** in which the vector can be propagated. In this chapter the various types of host cell will be described first, followed by vector systems and methods for getting DNA into cells.

Table 5.1. *Types of host cell used for genetic engineering*

Major group	Prokaryotic/ eukaryotic	Type	Examples
Bacteria	Prokaryotic	Gram −ve	*Escherichia coli*
		Gram +ve	*Bacillus subtilis*
			Streptomyces spp.
Fungi	Eukaryotic	Microbial	*Saccharomyces cerevisiae*
		Filamentous	*Aspergillus nidulans*
Plants	Eukaryotic	Protoplasts	Various types
		Intact cells	Various types
		Whole organisms	Various types
Animals	Eukaryotic	Insect cells	*Drosophila melanogaster*
		Mammalian cells	Various types
		Oocytes	Various types
		Whole organism	Various types

Note: Plant and animal cells may be subjected to manipulation either in tissue culture or as cells in the whole organisms. In some cases cells in culture may be hybrids formed by fusing cells of different species.

5.1 Host cell types

The type of host cell used for a particular application will depend mainly on the purpose of the cloning experiment. If the aim is to isolate a gene for structural analysis, the requirements may call for a simple system that is easy to use. If the aim is to express the genetic information in a higher eukaryote such as a plant, a more specific system will be required. These two aims are not necessarily mutually exclusive; often a simple **primary** host is used to isolate a sequence that is then introduced into a more complex system for expression. The main types of host cell are shown in Table 5.1, and are described below.

5.1.1 Prokaryotic hosts

An ideal host cell should be easy to handle and propagate, should be available as a wide variety of genetically defined strains, and should accept a range of vectors. The bacterium *Escherichia coli* fulfils these requirements, and is used in many cloning protocols. *E. coli* has been studied in great detail, and many different strains were isolated by microbial geneticists as they investigated the genetic mechanisms of this prokaryotic organism. Such studies provided the essential background information on which genetic engineering is based.

E. coli is a Gram-negative bacterium with a single chromosome packed into a compact structure known as the **nucleoid**. Genome size is some 4.6×10^6 base-pairs, and the complete sequence is now known. The processes of gene expression (transcription and translation) are coupled, with the newly synthesised mRNA being immediately available for translation. There is no post-transcriptional modification of the primary transcript as is commonly found in eukaryotic cells. *E. coli* can therefore be considered as one of the simplest host cells. Much of the gene cloning that is carried out routinely in laboratories involves the use of *E. coli* hosts, with many genetically different strains available for specific applications.

In addition to *E. coli*, other bacteria may be used as hosts for gene cloning experiments, with examples including species of *Bacillus*, *Pseudomonas* and *Streptomyces*. There are, however, certain drawbacks with most of these. Often there are fewer suitable vectors available for use in such cells than is the case for *E. coli*, and getting recombinant DNA into the cell can cause problems. This is particularly troublesome when primary cloning experiments are envisaged, i.e. direct introduction of ligated recombinant DNA into the host cell. It is often more sensible to perform the initial cloning in *E. coli*, isolate the required sequence, and then introduce the purified DNA into the target host. Many of the drawbacks can be overcome by using this approach, particularly when vectors that can function in the target host and in *E. coli* (**shuttle vectors**) are used. Use of bacteria other than *E. coli* will not be discussed further in this book; details may be found in some of the texts mentioned in the section Suggestions for further reading.

5.1.2 Eukaryotic hosts

One disadvantage of using an organism such as *E. coli* as a host for cloning is that it is a prokaryote, and therefore lacks the membrane-bound nucleus (and other organelles) found in eukaryotic cells. This means that certain eukaryotic genes may not function in *E. coli* as they would in their normal environment, which can hamper their isolation by selection mechanisms that depend on gene expression. Also, if the production of a eukaryotic protein is the desired outcome of a cloning experiment, it may not be easy to ensure that a prokaryotic host produces a fully functional protein.

Eukaryotic cells range from microbes, such as yeast and algae, to cells from complex multicellular organisms, such as ourselves. The microbial cells have many of the characteristics of bacteria with regard to ease of growth and availability of mutants. Higher eukaryotes present a different set of problems to the genetic engineer, many of which require specialised solutions. Often the aim of a gene manipulation experiment in a higher plant or animal is to alter the genetic makeup of the organism by creating a **transgenic**, rather than to isolate a gene for further analysis or to produce large amounts of a particular protein. Transgenesis is discussed further in Chapter 12.

The yeast *Saccharomyces cerevisiae* is one of the most commonly used eukaryotic microbes in genetic engineering. It has been used for centuries in the production of bread and beer, and has been studied extensively. The organism is amenable to classical genetic analysis, and a range of mutant cell types is available. In terms of genome complexity, *S. cerevisiae* has about 3.5 times more DNA than *E. coli*. The complete genome sequence is now known. Other fungi that may be used for gene cloning experiments include *Aspergillus nidulans* and *Neurospora crassa*.

Plant and animal cells may also be used as hosts for gene manipulation experiments. Unicellular algae such as *Chlamydomonas reinhardtii* have all the advantages of microorganisms plus the structural and functional organisation of plant cells, and their use in genetic manipulation will increase as they become more widely studied. Other plant cells (and animal cells) are usually grown as cell cultures, which are much easier to manipulate than cells in a whole organism. Some aspects of genetic engineering in plant and animal cells are discussed in the final section of this book.

5.2 Plasmid vectors for use in *E. coli*

There are certain essential features that vectors must possess. Ideally, they should be fairly small DNA molecules, to facilitate isolation and handling. There must be an **origin of replication**, so that their DNA can be copied and thus maintained in the cell population as the host organism grows and divides. It is desirable to have some sort of **selectable marker** that will enable the vector to be detected, and the vector must also have at least one unique restriction endonuclease recognition site, to enable DNA to be inserted during the production of recombinants. Plasmids have these features, and are extensively used as vectors in cloning experiments. Some features of plasmid vectors are described below.

5.2.1 What are plasmids?

Many types of plasmid are found in nature, in bacteria and some yeasts. They are circular DNA molecules, relatively small when compared to the host cell chromosome, that are maintained mostly in an extrachromosomal state. Although plasmids are generally dispensable (i.e. not essential for cell growth and division), they often confer traits (such as antibiotic resistance) on the host organism, which can be a selective advantage under certain conditions. The antibiotic resistance genes encoded by plasmid DNA (pDNA) are often used in the construction of vectors for genetic engineering, as they provide a convenient means of selecting cells containing the plasmid.

Plasmids can be classified into two groups, termed **conjugative** and **non-conjugative**. Conjugative plasmids can mediate their own transfer between bacteria by the process of conjugation, which requires functions specified by the *tra* (transfer) and *mob* (mobilising) regions carried on the plasmid. Non-conjugative plasmids are not self-transmissible, but may be mobilised by a conjugation-proficient plasmid if their *mob* region is functional. A further classification is based on the number of copies of the plasmid found in the host cell, a feature known as the **copy number**. Low copy number plasmids tend to exhibit **stringent** control of DNA replication, with replication of the pDNA closely tied to host cell chromosomal DNA replication. High copy number plasmids are termed **relaxed** plasmids, with DNA replication not dependent on host cell chromosomal DNA replication. In general terms, conjugative plasmids are large, show stringent control of DNA replication, and are present at low copy numbers, whilst non-conjugative plasmids are small, show relaxed DNA replication and are present at high copy numbers. Some examples of plasmids are shown in Table 5.2.

5.2.2 Basic cloning plasmids

For genetic engineering, naturally occurring plasmids have been extensively modified to produce vectors that have the desired characteristics. In naming plasmids, p is used to designate plasmid, and this is usually followed by the initials of the worker(s) who isolated or constructed the plasmid. Numbers may be used to classify the particular isolate. An important plasmid in the history of gene manipulation is pBR322, which was developed by Francisco Bolivar and his colleagues. Construction of pBR322 involved a series of manipulations to get the right pieces of DNA together, with the final result containing DNA from three sources. The plasmid has all the features of a good vector,

Table 5.2. *Properties of some naturally occurring plasmids*

Plasmid	Size (kb)	Conjugative	Copy number	Selectable markers
ColE1	7.0	No	10–15	$E1^{imm}$
RSF1030	9.3	No	20–40	Ap^r
CloDF13	10.0	No	10	$DF13^{imm}$
pSC101	9.7	No	1–2	Tc^r
RK6	42	Yes	10–40	Ap^r, Sm^r
F	103	Yes	1–2	—
R1	108	Yes	1–2	Ap^r, Cm^r, Sm^r, Sn^r, Km^r
RK2	56.4	Yes	3–5	Ap^r, Km^r, Tc^r

Note: Antibiotic abbreviations are as follows: Ap, ampicillin; Cm, chloramphenicol; Km, kanamycin; Sm, streptomycin; Sn, sulphonamide; Tc, tetracycline. $E1^{imm}$ and $DF13^{imm}$ represent immunity to the homologous but not to the heterologous colicin. Thus plasmid ColE1 is resistant to the effects of its own colicin (E1), but not to colicin DF13. Copy number is the number of plasmids per chromosome equivalent. ^{r}, resistance.
Source: After Winnacker (1987), *From Genes to Clones*, VCH. Data from Helsinki (1979), *Critical Reviews in Biochemistry* **7**, 83–101, copyright (1979) CRC Press, Inc., Boca Raton, Florida; Kahn *et al.* (1979), *Methods in Enzymology* **68**, 268–280, copyright (1979) Academic Press; Thomas (1981), *Plasmid* **5**, 10–19, copyright (1981) Academic Press. Reproduced with permission.

such as low molecular weight, antibiotic resistance genes, an origin of replication, and several single-cut restriction endonuclease recognition sites. A map of pBR322 is shown in Fig. 5.1.

There are several plasmids in the pBR series, each with slightly different features. In addition, the pBR series has been the basis for the development of many more plasmid vectors with additional features. In particular, one additional pBR322-based plasmid is worthy of inclusion, as it is still widely used and has some advantages over its progenitor. The plasmid is pAT153, which is a **deletion derivative** of pBR322 (see Fig. 5.1). The plasmid was isolated by removal of two fragments of DNA from pBR322, using the restriction enzyme *Hae*II. The amount of DNA removed was small (705 base-pairs), but the effect was to increase the copy number some threefold, and to remove sequences necessary for mobilisation. Thus pAT153 is, in some respects, a 'better' vector than pBR322, as it is present as more copies per cell and has a greater degree of biological containment because it is not mobilisable.

The presence of two antibiotic resistance genes (Ap^r and Tc^r) enables selection for cells harbouring the plasmid, as such cells will be resistant to both ampicillin and tetracycline. An added advantage is that the unique restriction

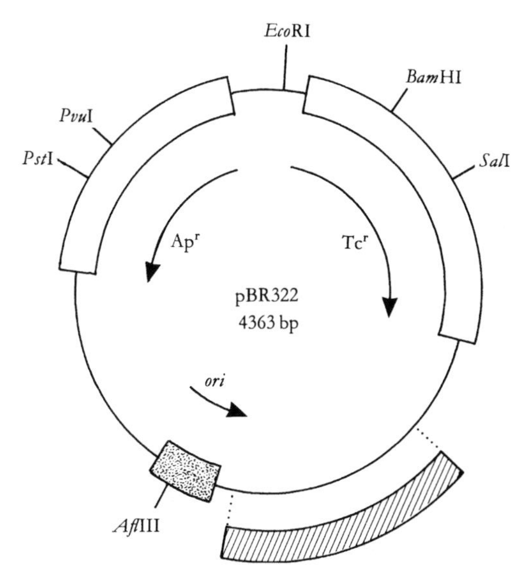

Fig. 5.1. Map of plasmid pBR322. Important regions indicated are the genes for ampicillin and tetracycline resistance (Ap^r and Tc^r) and the origin of replication (*ori*). Some unique restriction sites are given. The hatched region shows the two fragments that were removed from pBR322 to generate pAT153.

sites within the antibiotic resistance genes permit selection of cloned DNA by what is known as **insertional inactivation**, where the inserted DNA interrupts the coding sequence of the resistance gene and so alters the phenotype of the cell carrying the recombinant. This is discussed further in Section 8.1.2.

5.2.3 Slightly more exotic plasmid vectors

Although plasmids pBR322 and pAT153 are still often used for many applications in gene cloning, there are situations where other plasmid vectors may be more suitable. Generally, these have been constructed so that they have particular characteristics not found in the simpler vectors, such as a wider range of restriction sites for cloning DNA fragments. They may contain specific promoters for the expression of inserted genes, or they may offer other advantages such as direct selection for recombinants. Despite these advantages, the well-tried vectors such as pBR322 and pAT153 are often more than sufficient for the experimental procedure that is being performed.

One series of plasmid vectors that has proved popular is the pUC family. These plasmids have a region that contains several unique restriction endonuclease sites in a short stretch of the DNA. This region is known as a

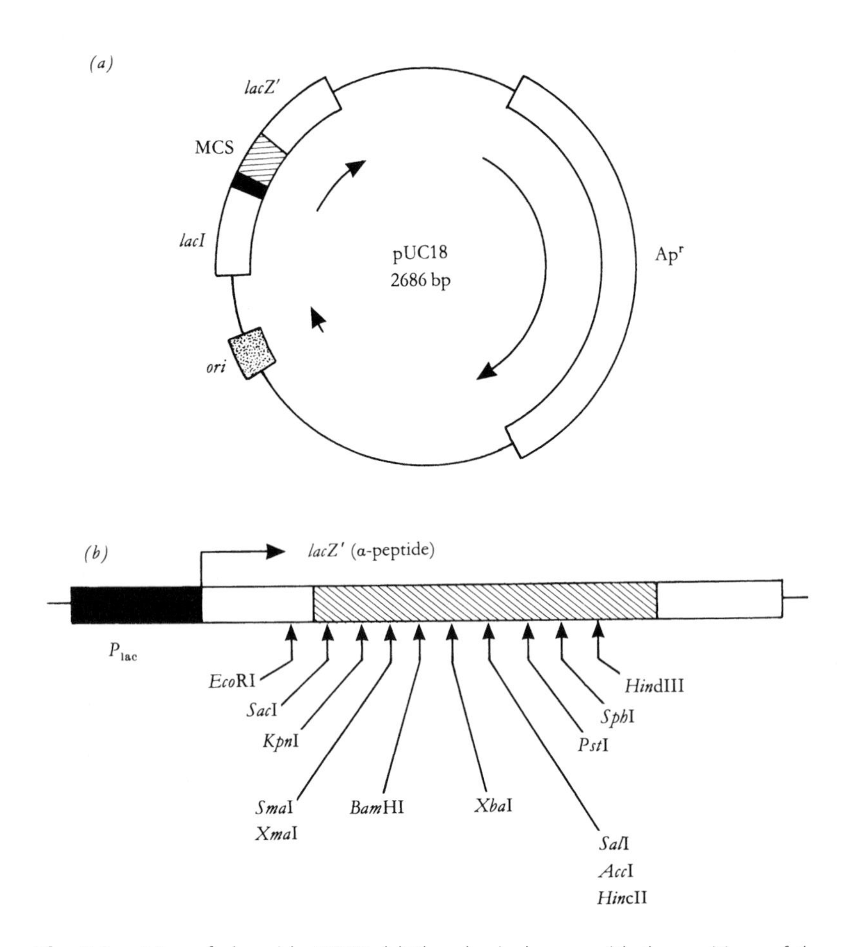

Fig. 5.2. Map of plasmid pUC18. (*a*) The physical map, with the positions of the origin of replication (*ori*) and the ampicillin resistance gene (Ap^r) indicated. The *lacI* gene (lac repressor), multiple cloning site (MCS) or polylinker and the *lacZ'* gene (α-peptide fragment of β-galactosidase) are also shown. (*b*) The polylinker region. This has multiple restriction sites immediately downstream from the lac promoter (P_{lac}). The in-frame insert used to create the MCS is hatched. Plasmid pUC19 is identical with pUC18 apart from the orientation of the polylinker region, which is reversed.

polylinker or **multiple cloning site** (MCS), and is useful because of the choice of site available for insertion of DNA fragments during recombinant production. A map of one of the pUC vectors, with the restriction sites in its polylinker region, is shown in Fig. 5.2. In addition to the multiple cloning sites in the polylinker region, the pUC plasmids have a region of the β-galactosidase gene that codes for what is known as the α-peptide. This sequence contains

Table 5.3. *Some commercially available plasmid vectors*

Vector	Features	Applications	Supplier
pBR322	Ap^r Tc^r Single cloning sites	General cloning & sub-cloning in *E. coli*	Various
pAT153	Ap^r Tc^r Single cloning sites	General cloning & sub-cloning in *E. coli*	Various
pGEM™-3Z	Ap^r MCS SP6/T7 promoters *lacZ* α-peptide	General cloning & *in vitro* transcription in *E. coli* and mammalian cells	Promega
pCI™	Ap^r MCS T7 promoter CMV enhancer/promoter	Expression of genes in mammalian cells	Promega
pCMV-Script™	Neo^r Large MCS CMV enhancer/promoter	Expression of genes in mammalian cells Cloning of PCR products	Stratagene

Note: There are hundreds of variants available from many different suppliers. A good source of information is the supplier's catalogue or website. Ap^r–ampicillin resistance; Tc^r–tetracycline resistance; Neo^r–neomycin resistance (selection using kanamycin in bacteria, G418 in mammalian cells); MCS – multiple cloning site; SP6/T7 are promoters for *in vitro* transcription; *lacZ* – β-galactosidase gene; CMV – human cytomegalovirus. The terms pGEM, pCI and pCMV-Script are trademarks of the suppliers.

the polylinker region, and insertion of a DNA fragment into one of the cloning sites results in a non-functional α-peptide. This forms the basis for a powerful direct recombinant screening method using the chromogenic substrate X-gal, as outlined in Sections 8.1.1 and 8.1.2.

Over the past few years, many different types of plasmid vectors have been derived from the basic cloning plasmids. Today there are many different plasmids available for specific purposes, often from commercial sources. These vectors are sometimes provided as part of a 'cloning kit' that contains all the essential components to conduct a cloning experiment. This has made the technology much more accessible to a greater number of scientists, although it has not yet become totally foolproof! Some commercially available plasmids are listed in Table 5.3.

Although plasmid vectors have many useful properties and are essential for gene manipulation, they do have a number of disadvantages. One of the major drawbacks is the size of DNA fragment that can be inserted into plasmids, the maximum being around 5 kb of DNA before cloning efficiency or insert stability are affected. In many cases this is not a problem, but in some applications it is important to maximise the size of fragments that may be cloned. Such a case is the generation of a **genomic library**, in which all the sequences present in the genome of an organism are represented. For this type of approach, vectors that can accept larger pieces of DNA are required. Examples of suitable vectors are those based on bacteriophage lambda (λ); these are considered in the next section.

5.3 Bacteriophage vectors for use in *E. coli*

Although bacteriophage-based vectors are in many ways more specialised than plasmid vectors, they fulfil essentially the same function, i.e. they act as carrier molecules for fragments of DNA. Two types of bacteriophage (λ and M13) have been extensively developed for cloning purposes; these will be described to illustrate the features of bacteriophages and the vectors derived from them.

5.3.1 What are bacteriophages?

In the 1940s Max Delbrück, and the 'Phage Group' that he brought into existence, laid the foundations of modern molecular biology by studying bacteriophages. These are literally 'eaters of bacteria', and are viruses that are dependent on bacteria for their propagation. The term bacteriophage is often shortened to **phage**, and can be used to describe either one or many particles of the same type. Thus we might say that a test tube contained one λ phage or 2×10^6 λ phage particles. The plural term **phages** is used when different types of phage are being considered; we therefore talk of T4, M13 and λ as being phages.

Structurally, phages fall into three main groups: (i) tailless, (ii) head with tail and (iii) filamentous. The genetic material may be single- or double-stranded DNA or RNA, with double-stranded DNA (dsDNA) being found most often. In tailless and tailed phages the genome is encapsulated in an icosahedral protein shell called a **capsid** (sometimes known as a phage coat or head). In typical dsDNA phages, the genome makes up about 50% of the mass of the phage particle. Thus phages represent relatively simple systems when compared to bacte-

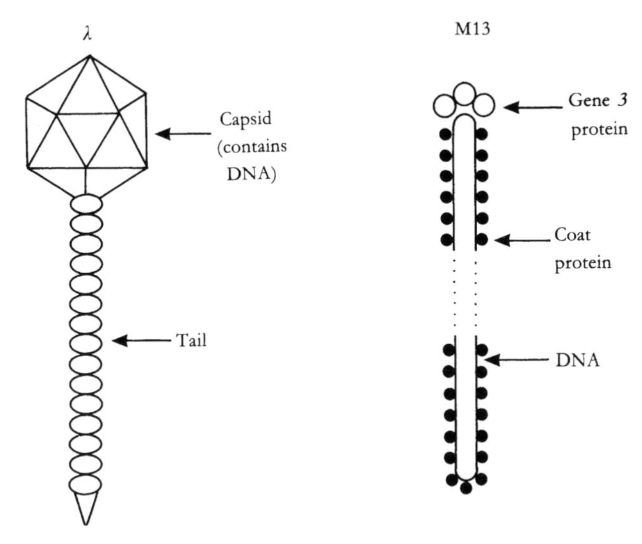

Fig. 5.3. Structure of bacteriophages λ and M13. Phage λ has a capsid or head which encloses the double-stranded DNA genome. The tail region is required for adsorption to the host cell. M13 has a simpler structure, with the single-stranded DNA genome being enclosed in a protein coat. The gene *3* product is important in both adsorption and extrusion of the phage. M13 is not drawn to scale; in reality it is a long thin structure.

ria, and for this reason they have been extensively used as models for the study of gene expression. The structure of phages λ and M13 is shown in Fig. 5.3.

Phages may be classified as either **virulent** or **temperate**, depending on their life cycles. When a phage enters a bacterial cell it can produce more phage and kill the cell (this is called the **lytic** growth cycle), or it can integrate into the chromosome and remain in a quiescent state without killing the cell (this is the **lysogenic** cycle). Virulent phages are those that exhibit a lytic life cycle only. Temperate phages exhibit lysogenic life cycles, but most can also undergo the lytic response when conditions are suitable. The best-known example of a temperate phage is λ, which has been the subject of intense research effort and is now more or less fully characterised in terms of its structure and mode of action.

The genome of phage λ is 48.5 kb in length, and encodes some 46 genes (Fig. 5.4). The entire genome has been sequenced (this was the first major sequencing project to be completed, and represents one of the milestones of molecular genetics), and all the regulatory sites are known. At the ends of the linear genome there are short (12 bp) single-stranded regions that are complementary. These act as cohesive or 'sticky' ends, which enable circularisation of

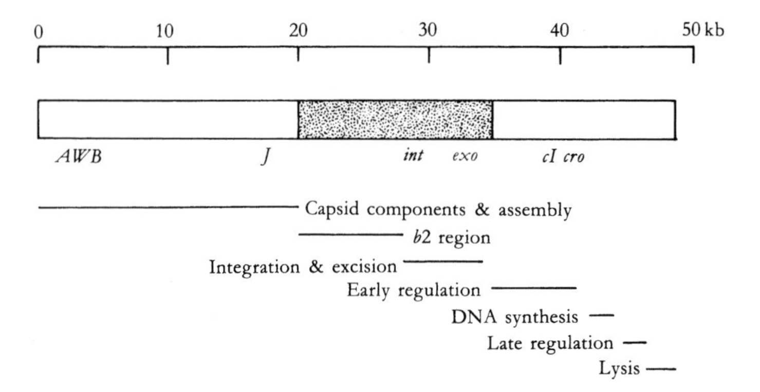

Fig. 5.4. Map of the phage λ genome. Some of the genes are indicated. Functional regions are shown by horizontal lines and annotated. The non-essential region that may be manipulated in vector construction is shaded.

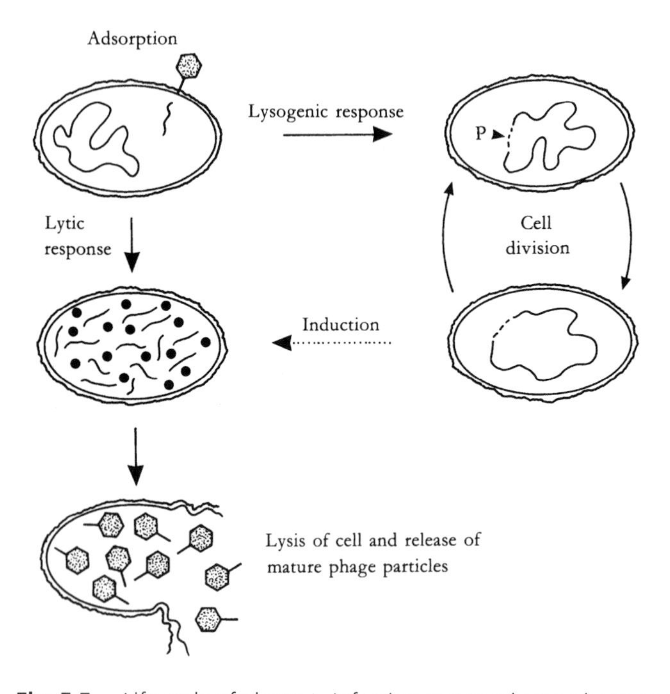

Fig. 5.5. Life cycle of phage λ. Infection occurs when a phage particle is adsorbed and the DNA injected into the host cell. In the lytic response, the phage takes over the host cell and produces copies of the phage genome and structural proteins. Mature phage particles are then assembled and released by lysis of the host cell. In the lysogenic response the phage DNA integrates into the host genome as a prophage (P), which can be maintained through successive cell divisions. The lytic response can be induced in a lysogenic bacterium in response to a stimulus such as ultraviolet light.

the genome following infection. The region of the genome that is generated by the association of the cohesive ends is known as the *cos* site.

Phage infection begins with **adsorption**, which involves the phage particle binding to receptors on the bacterial surface (Fig. 5.5). When the phage has adsorbed, the DNA is injected into the cell and the life cycle can begin. The genome circularises and the phage initiates either the lytic or lysogenic cycle, depending on a number of factors that include the nutritional and metabolic state of the host cell and the **multiplicity of infection** (m.o.i. – the ratio of phage to bacteria during adsorption). If the lysogenic cycle is initiated, the phage genome integrates into the host chromosome and is maintained as a **prophage**. It is then replicated with the chromosomal DNA and passed on to daughter cells in a stable form. If the lytic cycle is initiated, a complex sequence of transcriptional events essentially enables the phage to take over the host cell and produce multiple copies of the genome and the structural proteins. These components are then assembled or **packaged** into mature phage, which are released following lysis of the host cell.

To determine the number of bacteriophage present in a suspension, serial dilutions of the phage stock are mixed with an excess of indicator bacteria (m.o.i. is very low) and plated onto agar using a soft agar overlay. On incubation, the bacteria will grow to form what is termed a bacterial **lawn**. Phage that grow in this lawn will cause lysis of the cells that the phage infects, and as this growth spreads a cleared area or **plaque** will develop (Fig. 5.6). Plaques can then be counted to determine the number of **plaque forming units** (p.f.u.) in the stock suspension, and may be picked from the plate for further growth and analysis. Phage may be propagated in liquid culture by infecting a growing culture of the host cell and incubating until cell lysis is complete, the yield of phage particles depending on the multiplicity of infection and the stage in the bacterial growth cycle at which infection occurs.

The filamentous phage M13 differs from λ both structurally (Fig. 5.3) and in its life cycle. The M13 genome is a single-stranded circular DNA molecule 6407 bp in length. The phage will infect only *E. coli* that have F-pili (threadlike protein 'appendages' found on conjugation-proficient cells). When the DNA enters the cell, it is converted to a double-stranded molecule known as the **replicative form** or **RF**, which replicates until there are about 100 copies in the cell. At this point DNA replication becomes asymmetric, and single-stranded copies of the genome are produced and extruded from the cell as M13 particles. The bacterium is not lysed and remains viable during this process, although growth and division are slower than in non-infected cells.

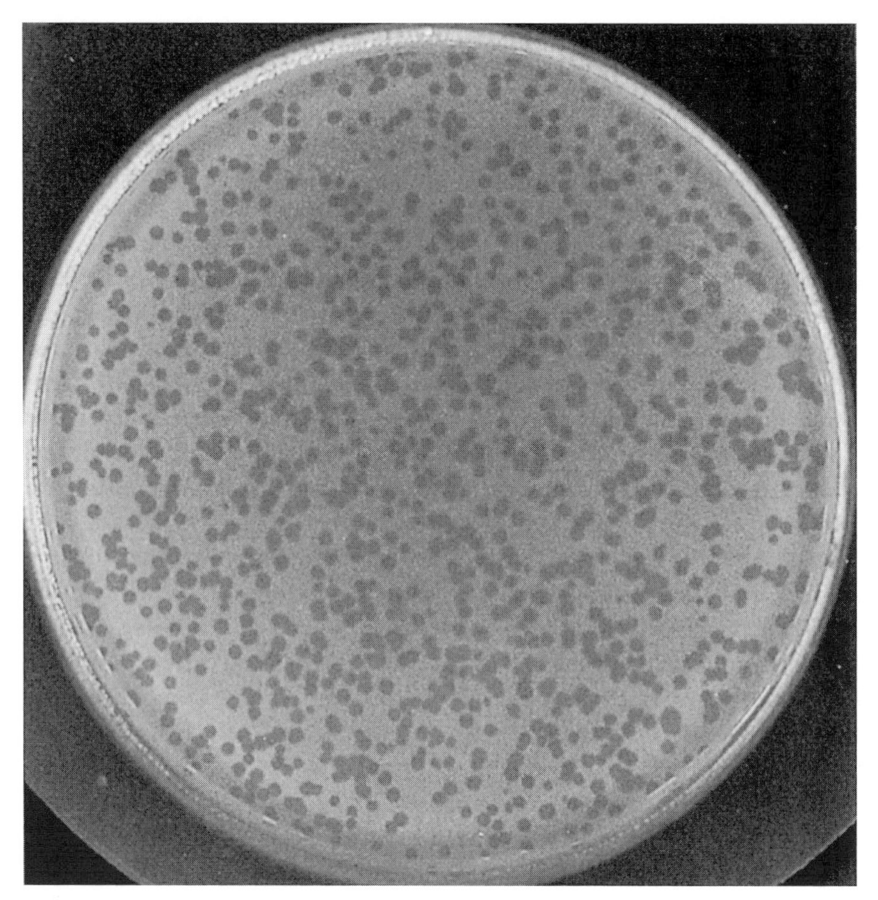

Fig. 5.6. Bacteriophage plaques. Particles of phage λ were mixed with a strain of *E. coli* and plated using a soft agar overlay. After overnight incubation the bacterial cells grow to form a lawn, in which regions of phage infection appear as cleared areas or plaques. (Photograph courtesy of Dr M. Stronach.)

5.3.2 Vectors based on bacteriophage λ

The utility of phage λ as a cloning vector depends on the fact that not all of the λ genome is essential for the phage to function. Thus there is scope for the introduction of exogenous DNA, although certain requirements have had to be met during the development of cloning vectors based on phage λ. Firstly, the arrangement of genes on the λ genome will determine which parts can be removed or replaced for the addition of exogenous DNA. It is fortunate that the central region of the λ genome (between positions 20 and 35 on the map shown in Fig. 5.4) is largely dispensable, so no complex rearrangement of the genome *in vitro* is required. The central region controls mainly the lysogenic properties of the phage, and much of this region can be deleted without impairing the functions required for the lytic infection cycle. Secondly, wild-type λ phage will generally have multiple recognition sites for the restriction enzymes commonly used in cloning procedures. This can be a major problem, as it limits the choice of sites for the insertion of DNA. In practice, it is relatively easy to select for phage that have reduced numbers of sites for particular restriction enzymes, and the technique of mutagenesis *in vitro* may be used to modify remaining sites that are not required. Thus it is possible to obtain phage that have the desired combination of restriction enzyme recognition sites.

One of the major drawbacks of λ vectors is that the capsid places a physical constraint on the amount of DNA that can be incorporated during phage assembly, which limits the size of exogenous DNA fragments that can be cloned. During packaging, viable phage particles can be produced from DNA that is between approximately 38 kb and 51 kb in length. Thus a wild-type phage genome could accommodate only around 2.5 kb of cloned DNA before becoming too large for viable phage production. This limitation has been minimised by careful construction of vectors to accept pieces of DNA that are close to the theoretical maximum for the particular construct. Such vectors fall into two main classes: (i) **insertion** vectors and (ii) **replacement** or **substitution** vectors. The difference between these two types of vector is outlined in Fig. 5.7.

As with plasmids, there is now a bewildering variety of λ vectors available for use in cloning experiments, each with slightly different characteristics. The choice of vector has to be made carefully, with aspects such as the size of DNA fragments to be cloned and the preferred selection/screening method being taken into account. To illustrate the structural characteristics of λ vectors, two insertion and two replacement vectors are described briefly. Although not as widely used as some other λ vectors today, these illustrate some of the important aspects of vector design. Functional aspects of some λ vectors are discussed in Chapter 8 when selection and screening methods are considered.

Insertion vectors have a single recognition site for one or more restriction enzymes, which enables DNA fragments to be inserted into the λ genome. Examples of insertion vectors include λgt10 and Charon 16A. The latter is one of a series of vectors named after the ferryman of Greek mythology, who conveyed the spirits of the dead across the river Styx – a rather apt example of what we might call 'bacteriophage culture'! These two insertion vectors are

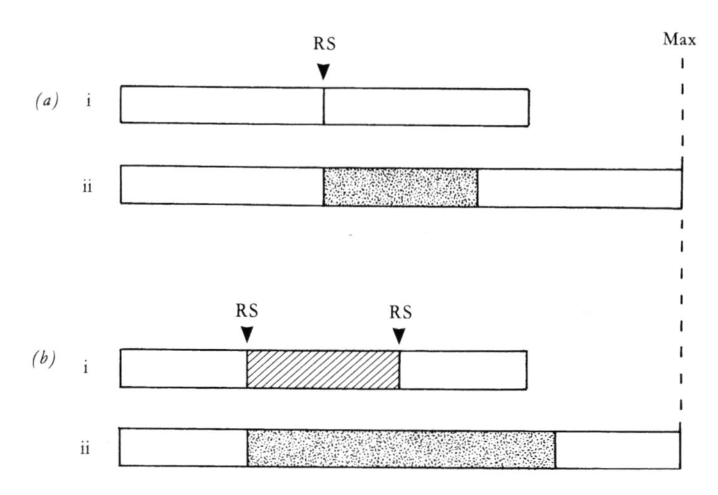

Fig. 5.7. Insertion and replacement phage vectors. (*a*) An insertion vector is shown in part i. Such vectors have a single restriction site (RS). To generate a recombinant DNA is inserted into this site. The size of fragment that may be cloned is therefore determined by the difference between the vector size and the maximum packagable fragment size (Max). Insert DNA is shaded in part ii. (*b*) A replacement vector is shown in part i. These vectors have two restriction sites (RS) which flank a region known as the stuffer fragment (hatched). Thus a section of the phage genome is replaced during cloning into this site, as shown in part ii. This approach enables larger fragments to be cloned than is possible with insertion vectors.

illustrated in Fig. 5.8. Each has a single *Eco*RI site into which DNA can be inserted. In λgt10 (43.3 kb) this generates left and right 'arms' of 32.7 and 10.6 kb, respectively, which can in theory accept insert DNA fragments up to approximately 7.6 kb in length. The *Eco*RI site lies within the *cI* gene (λ repressor), and this forms the basis of a selection/screening method based on plaque formation and morphology (see Section 8.1.2). In Charon 16A (41.8 kb), the arms generated by *Eco*RI digestion are 19.9 kb (left arm) and 21.9 kb (right arm), and fragments of up to approximately 9 kb may be cloned. The *Eco*RI site in Charon 16A lies within the β-galactosidase gene (*lacZ*), which enables the detection of recombinants using X-gal (see Section 8.1.2).

Insertion vectors offer limited scope for cloning large pieces of DNA, and thus replacement vectors were developed in which a central 'stuffer' fragment is removed and replaced with the insert DNA. Two examples of λ replacement vectors are EMBL4 and Charon 40 (Fig. 5.9). EMBL4 (41.9 kb) has a central 13.2 kb stuffer fragment flanked by inverted polylinker sequences containing sites for the restriction enzymes *Eco*RI, *Bam*HI and *Sal*I. Two *Sal*I sites

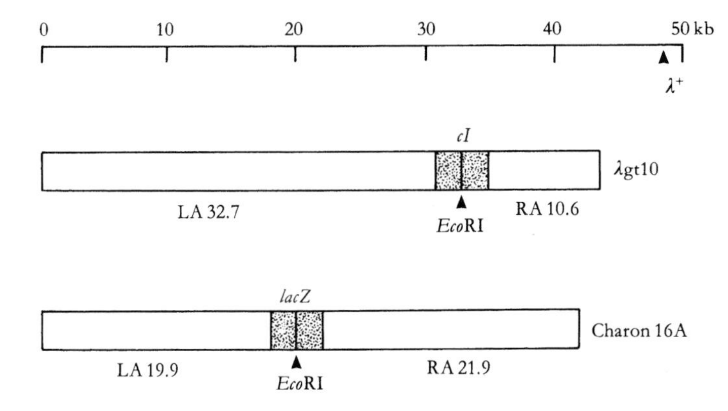

Fig. 5.8. Bacteriophage λ insertion vectors λgt10 and Charon 16A. The *cI* and *lacZ* genes in λgt10 and Charon 16A, respectively, are shaded. Within these genes there is an *Eco*RI site for cloning into. The lengths of the left and right arms (LA and RA, in kb) are given. The size of the wild-type λ genome is marked on the scale bar as λ^+. Redrawn from Winnacker (1987), *From Genes to Clones*, VCH. Reproduced with permission.

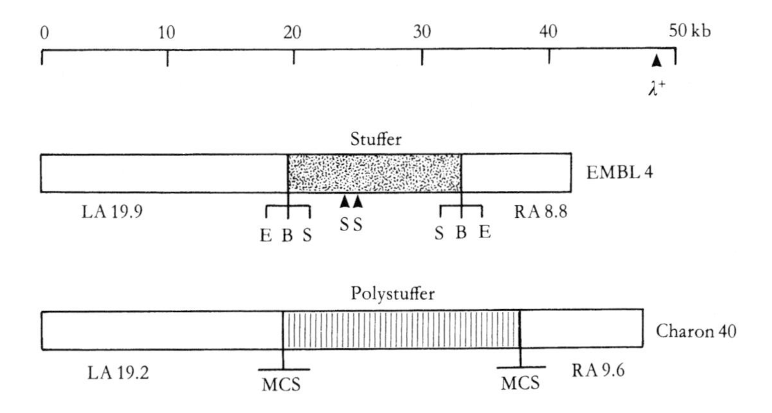

Fig. 5.9. Bacteriophage λ replacement vectors EMBL4 and Charon 40. The stuffer fragment in EMBL4 is 13.2 kb, and is flanked by inverted polylinkers containing the sites for *Eco*RI (E), *Bam*HI (B) and *Sal*I (S). In Charon 40 the polystuffer is composed of short repeated regions that are cleaved by *Nae*I. The multiple cloning site (MCS) in Charon 40 carries a wider range of restriction sites than that in EMBL4.

are also present in the stuffer fragment. DNA may be inserted into any of the cloning sites, the choice depending on the method of preparation of the fragments. Often a partial *Sau*3A or *Mbo*I digest is used in the preparation of a genomic library (see Section 6.3.2), which enables insertion into the *Bam*HI site. Such inserts may be released from the recombinant by digestion with

*Eco*RI. During preparation of the vector for cloning, the *Bam*HI digestion (which generates sticky ends for accepting the insert DNA) is often followed by a *Sal*I digestion. This cleaves the stuffer fragment at the two internal *Sal*I sites and also releases short *Bam*HI/*Sal*I fragments from the polylinker region. This is helpful because it prevents the stuffer fragment from re-annealing with the left and right arms and generating a viable phage that is non-recombinant.

DNA fragments between approximately 9 and 22 kb may be cloned in EMBL4, the lower limit representing the minimum size required to form viable phage particles (left arm + insert + right arm must be greater than 38 kb) and the upper the maximum packagable size of around 51 kb. These size constraints can act as a useful initial selection method for recombinants, although an additional genetic selection mechanism can be employed with EMBL4 (the Spi^- phenotype, see Section 8.1.4).

Charon 40 is a replacement vector in which the stuffer fragment is composed of multiple repeats of a short piece of DNA. This is known as a **polystuffer**, and it has the advantage that the restriction enzyme *Nae*I will cut the polystuffer into its component parts. This enables efficient removal of the polystuffer during vector preparation, and most of the surviving phage will be recombinant. The polystuffer is flanked by polylinkers with a more extensive range of restriction sites than found in EMBL4, which increases the choice of restriction enzymes that may be used to prepare the insert DNA. The size range of fragments that may be cloned in Charon 40 is similar to that for EMBL4.

5.3.3 Vectors based on bacteriophage M13

Two aspects of M13 infection are of value to the genetic engineer. Firstly, the RF is essentially similar to a plasmid, and can be isolated and manipulated using the same techniques. A second advantage is that the single-stranded DNA produced during the infection is useful in techniques such as DNA sequencing by the dideoxy method (see Section 3.6.2). This aspect alone made M13 immediately attractive as a potential vector.

Unlike phage λ, M13 does not have any non-essential genes. The 6407 bp genome is also used very efficiently in that most of it is taken up by gene sequences, so that the only part available for manipulation is a 507 bp intergenic region. This has been used to construct the M13mp series of vectors, by inserting a polylinker/*lacZ* α-peptide sequence into this region (Fig. 5.10). This enables the X-gal screening system to be used for the detection of recombinants, as is the case with the pUC plasmids. When M13 is grown on

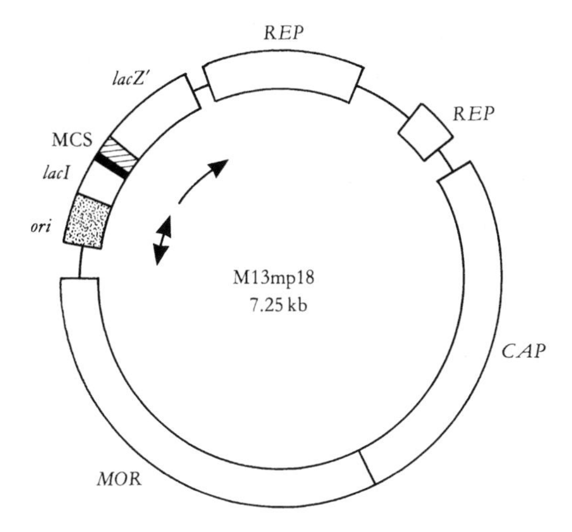

Fig. 5.10. Map of the filamentous phage vector M13mp18. The double-stranded replicative form is shown. The polylinker region (MCS) is the same as that found in plasmid pUC18 (Fig. 5.2). Genes in the *REP* region encode proteins that are important for DNA replication. The *CAP* and *MOR* regions contain genes that specify functions associated with capsid formation and phage morphogenesis, respectively. The vector M13mp19 is identical except for the orientation of the polylinker region.

a bacterial lawn, 'plaques' appear due to the reduction in growth of the host cells (which are not lysed), and these may be picked for further analysis.

A second disadvantage of M13 vectors is the fact that they do not function efficiently when long DNA fragments are inserted into the vector. Although in theory there should be no limit to the size of clonable fragments, as the capsid structure is determined by the genome size (unlike phage λ), there is a marked reduction in cloning efficiency with fragments longer than about 1.5 kb. In practice this was not a major problem, as the main use of the early M13 vectors was in sub-cloning small DNA fragments for sequencing. In this application single-stranded DNA production, coupled with ease of purification of the DNA from the cell culture, outweighs any size limitation, although this has also been alleviated by the construction of hybrid plasmid/M13 vectors (see Section 5.4.1).

5.4 Other vectors

So far I have concentrated on what we might call 'basic' plasmid and bacteriophage vectors for use in *E. coli* hosts. Although these vectors still represent a major part of the technology of gene manipulation, there has been continued development of more sophisticated bacterial vectors, as well as vectors for other organisms. One driving force in this has been the need to clone and analyse ever larger pieces of DNA, as the emphasis in molecular biology has shifted towards the analysis of genomes rather than simply genes in isolation. In this section I examine the features of some additional bacterial vectors, and some vectors for use in other organisms.

5.4.1 Hybrid plasmid/phage vectors

One feature of phage vectors is that the technique of packaging *in vitro* (see Section 5.5.2 for details) is sequence-independent, apart from the requirement of having the *cos* sites separated by DNA of packagable size (38–51 kb). This has been exploited in the construction of vectors that are made up of plasmid sequences joined to the *cos* sites of phage λ. Such vectors are known as **cosmids**. They are small (4–6 kb) and can therefore accommodate cloned DNA fragments up to some 47 kb in length. As they lack phage genes, they behave as plasmids when introduced into *E. coli* by the packaging/infection mechanism of λ. Cosmid vectors, therefore, offer an apparently ideal system – a highly efficient and specific method of introducing the recombinant DNA into the host cell, and a cloning capacity some twofold greater than the best λ replacement vectors. However, they are not without disadvantages, and often the gains of using cosmids instead of phage vectors are offset by losses in terms of ease of use and further processing of cloned sequences.

Hybrid plasmid/phage vectors in which the phage functions are expressed and utilised in some way are known as **phagemids** (sometimes also called **phasmids**). One such series of vectors is the λZAP family, produced by Stratagene. Features of these include the potential to excise cloned DNA fragments *in vivo* as part of a plasmid. This automatic excision is useful in that it removes the need to sub-clone inserts from λ into plasmid vectors for further manipulations.

Hybrid plasmid/phage vectors have been developed to overcome the size limitation of the M13 cloning system, and are now widely used for applications such as DNA sequencing and the production of probes for use in hybridisation studies. These vectors are essentially plasmids, which contain the f1 (M13) phage origin of replication. When cells containing the plasmid are superinfected with phage, they produce single-stranded copies of the plasmid DNA and secrete these into the medium as M13-like particles. Vectors such as pEMBL9 or pBluescript can accept DNA fragments of up to

Table 5.4. *Some commercially available bacteriophage-based vectors*

Vector	Features	Applications	Supplier
λGT11	λ insertion vector Insert capacity 7.2 kbp *lacZ* gene	cDNA library construction Expression of inserts	Various
λEMBL3/4	λ replacement vectors Insert capacity 9–23 kbp	Genomic library construction	Various
λZAP Express™	λ-based insertion vector Capacity of 12 kbp *In vivo* excision of inserts Expression of inserts	cDNA library construction Also genomic/PCR cloning	Stratagene
λFIX ™II	λ-based replacement vector, capacity 9–23 kbp Spi^+/P2 selection system to reduce non-recombinant background	Genomic library construction	Stratagene
pBluescript™II	Phagemid vector Produces single-stranded DNA	*In vitro* transcription DNA sequencing	Stratagene
SuperCos™	Cosmid vector with Ap^r and Neo^r markers, plus T3, T7 and SV40 promoters Capacity 30–42 kbp	Generation of cosmid-based genomic DNA libraries T3/7 promoters allow end-specific transcripts to be generated for chromosome walking techniques	Stratagene

Note: As with plasmid vectors, there are many variants available from a range of different suppliers. A good source of information is the supplier's catalogue or website. Ap^r–ampicillin resistance; Neo^r–neomycin resistance (selection using kanamycin in bacteria, G418 in mammalian cells); T3/7 are promoters for *in vitro* transcription; *lacZ* – β-galactosidase gene; SV40 – promoter for expression in eukaryotic cells. Terms marked TM are trademarks of Stratagene.

10 kb. Some commercially available vectors based on bacteriophages are listed in Table 5.4.

5.4.2 Vectors for use in eukaryotic cells

When eukaryotic host cells are considered, vector requirements become more complex. Bacteria are relatively simple in genetic terms, whereas

Table 5.5. *Some possible vectors for plant and animal cells*

Cell type	Vector type	Genome	Examples
Plant cells	Plasmid	DNA	Ti plasmids of *Agrobacterium tumefaciens*
	Viral	DNA	Cauliflower mosaic virus, Geminiviruses
		RNA	Tobacca mosaic virus
Animal cells	Plasmid	DNA	Various types of plasmid vector are available. Many are hybrid vectors containing part of the SV40 genome
	Viral	DNA	Baculoviruses Papilloma viruses Simian virus 40 (SV40) Vaccinia virus
	Viral	RNA	Retroviruses
	Transposon	DNA	P elements in *Drosophila melanogaster*

Note: In many cases the 'vectorology' associated with a particular group of potential vectors is not well advanced, and often well-tried vectors continue to be used and developed further for particular applications.

eukaryotic cells have multiple chromosomes that are held within the membrane-bound nucleus. Given the wide variety of eukaryotes, it is not surprising that vectors tend to be highly specialised and designed for specific purposes.

A range of vectors for use in yeast cells has been developed, with the choice of vector depending on the particular application. **Yeast episomal plasmids** (YEps) are based on the naturally occurring yeast 2 μm plasmid, and may replicate autonomously or integrate into a chromosomal location. **Yeast integrative plasmids** (YIps) are designed to integrate into the chromosome in a similar way to the YEp plasmids, and **yeast replicative plasmids** (YRps) remain as independent plasmids and do not integrate. Plasmids that contain sequences from around the centromeric region of chromosomes are known as **yeast centromere plasmids** (YCps), and these behave essentially as minichromosomes.

When dealing with higher eukaryotes that are multicellular, such as plants and animals, the problems of introducing recombinant DNA into the organism become slightly different than those which apply to microbial eukaryotes such as yeast. The aims of genetic engineering in higher eukaryotes are twofold: (i) to introduce recombinant DNA into plant and animal cells in tissue culture, for basic research on gene expression or for the production of useful proteins, and (ii) to alter the genetic makeup of the organism and produce a transgenic, in which all the cells will carry the genetic modification. The latter aim in particular can pose technical difficulties, as the recombinant DNA has to be introduced very early in development or in some sort of vector that will promote the spread of the recombinant sequence throughout the organism.

Vectors used for plant and animal cells may be introduced into cells directly by techniques such as those described in Section 5.5.3, or they may have a biological entry mechanism if based on viruses or other infectious agents such as *Agrobacteria*. Some examples of the types of system that have been used in the development of vectors for plant and animal cells are given in Table 5.5. The use of specific vectors is described further when considering the production of transgenics in Chapter 12.

5.4.3 Artificial chromosomes

The development of vectors for cloning very large pieces of DNA was essential to enable large genome sequencing projects to proceed at a reasonable rate, although genomes such as *Saccharomyces cerevisiae* have been sequenced mainly by using cosmid vectors to construct the genomic libraries required. However, even insert sizes of 40–50 kb are too small to cope with projects such as the Human Genome Project (see Chapter 9 for a more detailed treatment of genome analysis). The development of **yeast artificial chromosomes** (YACs; see Fig. 5.11) has enabled DNA fragments in the megabase range to be cloned, although there have been some problems of insert instability. YACs are the most sophisticated yeast vectors, and to date represent the largest capacity vectors available. They have centromeric and telomeric regions, and the recombinant DNA is therefore maintained essentially as a yeast chromosome.

A further development of artificial chromosome technology came with the construction of **bacterial artificial chromosomes** (BACs). These are based on the F plasmid, which is much larger than the standard plasmid cloning vectors, and therefore offers the potential of cloning larger fragments. BACs can accept inserts of around 300 kb, and many of the instability problems of YACs can be avoided by using the bacterial version. Much of the sequencing of the human genome has been accomplished using a library of BAC recombinants.

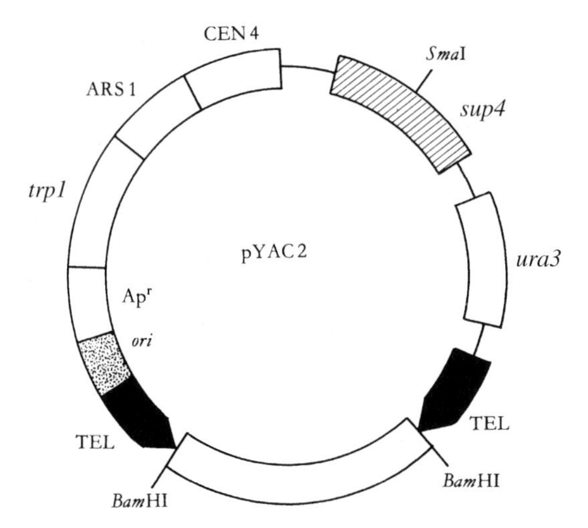

Fig. 5.11. Map of the yeast artificial chromosome vector pYAC2. This carries the origin of replication (*ori*; shaded) and ampicillin resistance gene (Apr) from pBR322, and yeast sequences for replication (ARS1) and chromosome structure (centromere, CEN4; and telomeres, TEL). The TEL sequences are separated by a fragment flanked by two *Bam*HI sites. The genes *trp1* and *ura3* may be used as selectable markers in yeast. The cloning site *Sma*I lies within the *sup4* gene (hatched). From Kingsman and Kingsman (1988), *Genetic Engineering*, Blackwell. Reproduced with permission.

5.5 Getting DNA into cells

Manipulation of vector and insert DNAs to produce recombinant molecules is carried out in the test-tube, and we are then faced with the task of getting the recombinant DNA into the host cell for propagation. The efficiency of this step is often a crucial factor in determining the success of a given cloning experiment, particularly when a large number of recombinants is required. The methods available depend on the type of host/vector system, and range from very simple procedures to much more complicated and esoteric ones. In this section I look at some of the methods available for getting recombinant DNA into host cells.

5.5.1 Transformation and transfection

The techniques of transformation and transfection represent the simplest methods available for getting recombinant DNA into cells. In the context of cloning in *E. coli* cells, transformation refers to the uptake of plasmid DNA, and transfection to the uptake of phage DNA. Transformation is also used more generally to describe uptake of any DNA by any cell, and can also be used in a different context when talking about a **growth transformation** such as occurs in the production of a cancerous cell.

Transformation in bacteria was first demonstrated in 1928 by Frederick Griffith, in his famous 'transforming principle' experiment that paved the way for the discoveries that eventually showed that genes were made of DNA. However, not all bacteria can be transformed easily, and it was not until the early 1970s that transformation was demonstrated in *E. coli*, the mainstay of gene manipulation technology. To effect transformation of *E. coli*, the cells need to be made **competent**. This is achieved by soaking the cells in an ice-cold solution of calcium chloride, which induces competence in a way that is still not fully understood. Transformation of competent cells is carried out by mixing the plasmid DNA with the cells, incubating on ice for 20–30 min, and then giving a brief heat shock (2 min at 42 °C is often used) which appears to enable the DNA to enter the cells. The transformed cells are usually incubated in a nutrient broth at 37 °C for 60–90 min to enable the plasmids to become established and permit phenotypic expression of their traits. The cells can then be plated out onto selective media for propagation of cells harbouring the plasmid.

Transformation is an inefficient process in that only a very small percentage of competent cells become transformed, representing uptake of a fraction of the plasmid DNA that is available. Thus the process can become the critical step in a cloning experiment where a large number of individual recombinants is required, or when the starting material is limiting. Despite these potential disadvantages, transformation is an essential technique, and with care can yield up to 10^9 transformed cells (**transformants**) per microgram of input DNA, although transformation frequencies of around 10^6 or 10^7 transformants per microgram are more often achieved in practice. Transfection is a similar process to transformation, the difference being that phage DNA is used instead of plasmid DNA. It is again a somewhat inefficient process, and it has largely been superseded by packaging *in vitro* for applications that require the introduction of phage DNA into *E. coli* cells.

5.5.2 Packaging phage DNA *in vitro*

During the lytic cycle of phage λ, the phage DNA is replicated to form what is known as a **concatemer**. This is a very long DNA molecule composed of

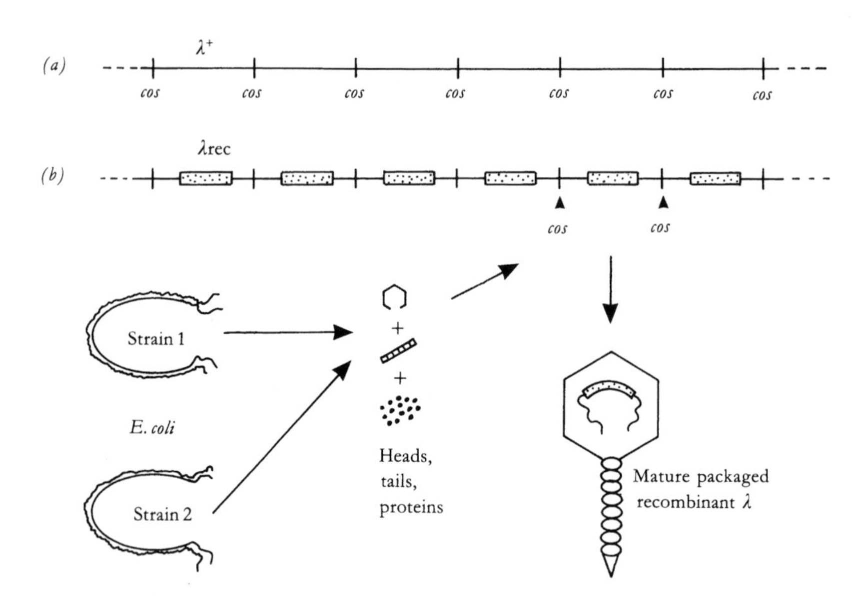

Fig. 5.12. Phage DNA and packaging. (*a*) A concatemeric DNA molecule composed of wild-type phage DNA (λ+). The individual genomes are joined at the cos sites. (*b*) Recombinant genomes (λrec) are shown being packaged *in vitro*. A mixed lysate from two bacterial strains supplies the head and tail precursors and the proteins required for the formation of mature λ particles. On adding this mixture to the concatemer, the DNA is cleaved at the *cos* sites (arrowed) and packaged into individual phage particles, each containing a recombinant genome.

many copies of the λ genome, linked together by the *cos* sites (Fig. 5.12(*a*)). When the phage particles are assembled the DNA is packaged into the capsid, which involves cutting the DNA at the *cos* sites using a phage-encoded endonuclease. Mature phage particles are thus produced, ready to be released on lysis of the cell, and capable of infecting other cells. This process normally occurs *in vivo*, the particular functions being encoded by the phage genes. However, it is possible to carry out the process in the test-tube, which enables recombinant DNA that is generated as a concatemer to be packaged into phage particles.

To enable packaging *in vitro*, the components of the λ capsid, and the endonuclease, must be available. In practice, two strains of bacteria are used to produce a lysate known as a **packaging extract**. Each strain is mutant in one function of phage morphogenesis, so that the packaging extracts will not work in isolation. When the two are mixed with the concatemeric recombinant DNA under suitable conditions, all the components are available and phage particles are produced. These particles can then be used to infect *E. coli* cells, which are plated out to obtain plaques. The process of packaging *in vitro* is summarised in Fig. 5.12(*b*).

5.5.3 Alternative DNA delivery methods

The methods available for introducing DNA into bacterial cells are not easily transferred to other cell types. The phage-specific packaging system is not available for other systems, and transformation by normal methods may prove impossible or too inefficient to be a realistic option. However, there are alternative methods for introducing DNA into cells. Often, these are more technically demanding and less efficient than the bacterial methods, but reliable results have been achieved in many situations where there appeared to be no hope of getting recombinant DNA molecules into the desired cell.

Most of the problems associated with getting DNA into non-bacterial cells have involved plant cells. Animal cells are relatively flimsy, and can be transformed readily. However, plant cells pose the problem of a rigid cell wall,

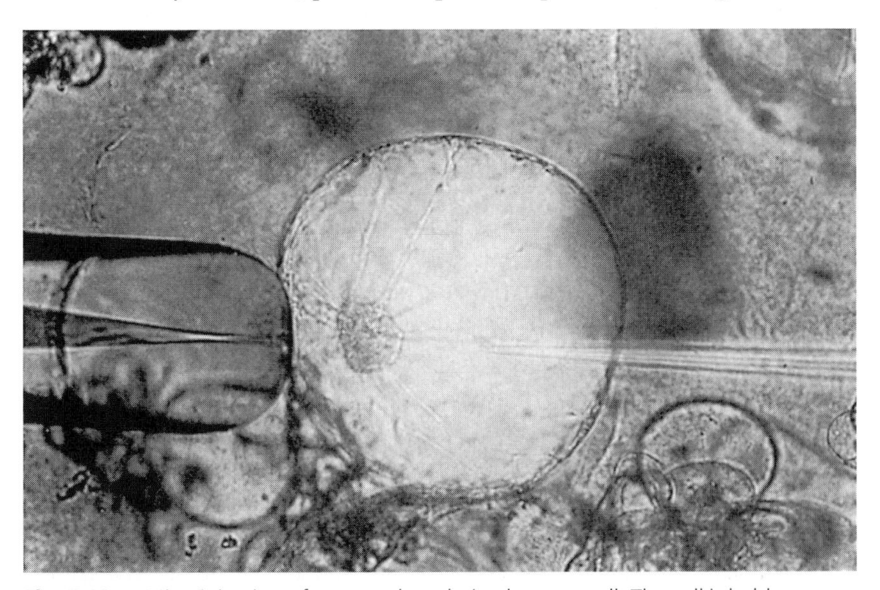

Fig. 5.13. Microinjection of a protoplast-derived potato cell. The cell is held on a glass capillary (on the left of the photograph) by gentle suction. The microinjection needle is made by drawing a heated glass capillary out to a fine point. Using a micromanipulator the needle has been inserted into the cell (on the right of the photograph), where its tip can be seen approaching the cell nucleus. (Photograph courtesy of Dr K. Ward.)

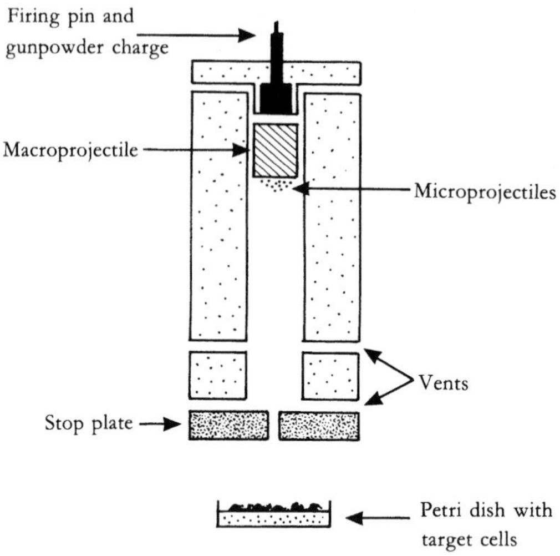

Fig. 5.14. Biolistic apparatus. The DNA is coated onto microprojectiles, which are accelerated by the macroprojectile on firing the gun. At the stop plate the macroprojectile is retained in the chamber and the microprojectiles carry on to the target tissue. Other versions of the apparatus, driven by compressed gas instead of a gunpowder charge, are available.

which is a barrier to DNA uptake. This can be alleviated by the production of **protoplasts**, in which the cell wall is removed enzymatically. The protoplasts can then be transformed using a technique such as **electroporation**, where an electrical pulse is used to create transient holes in the cell membrane, through which the DNA can pass. The protoplasts can then be regenerated. In addition to this application, protoplasts also have an important role to play in the generation of hybrid plant cells by fusing protoplasts together.

An alternative to transformation procedures is to introduce DNA into the cell by some sort of physical method. One way of doing this is to use a very fine needle and inject the DNA directly into the nucleus. This technique is called **microinjection** (Fig. 5.13), and has been used successfully with both plant and animal cells. The cell is held on a glass tube by mild suction and the needle used to pierce the membrane. The technique requires a mechanical micromanipulator and a microscope, and plenty of practice!

An ingenious and somewhat bizarre development has proved extremely useful in transformation of plant cells. The technique, which is called **biolistic** DNA delivery, involves literally shooting DNA into cells (Fig. 5.14). The DNA is used to coat microscopic tungsten particles known as **microprojec-**

tiles, which are then accelerated on a **macroprojectile** by firing a gunpowder charge, or by using compressed gas to drive the macroprojectile. At one end of the 'gun' there is a small aperture that stops the macroprojectile but allows the microprojectiles to pass through. When directed at cells, these microprojectiles carry the DNA into the cell and, in some cases, stable transformation will occur.

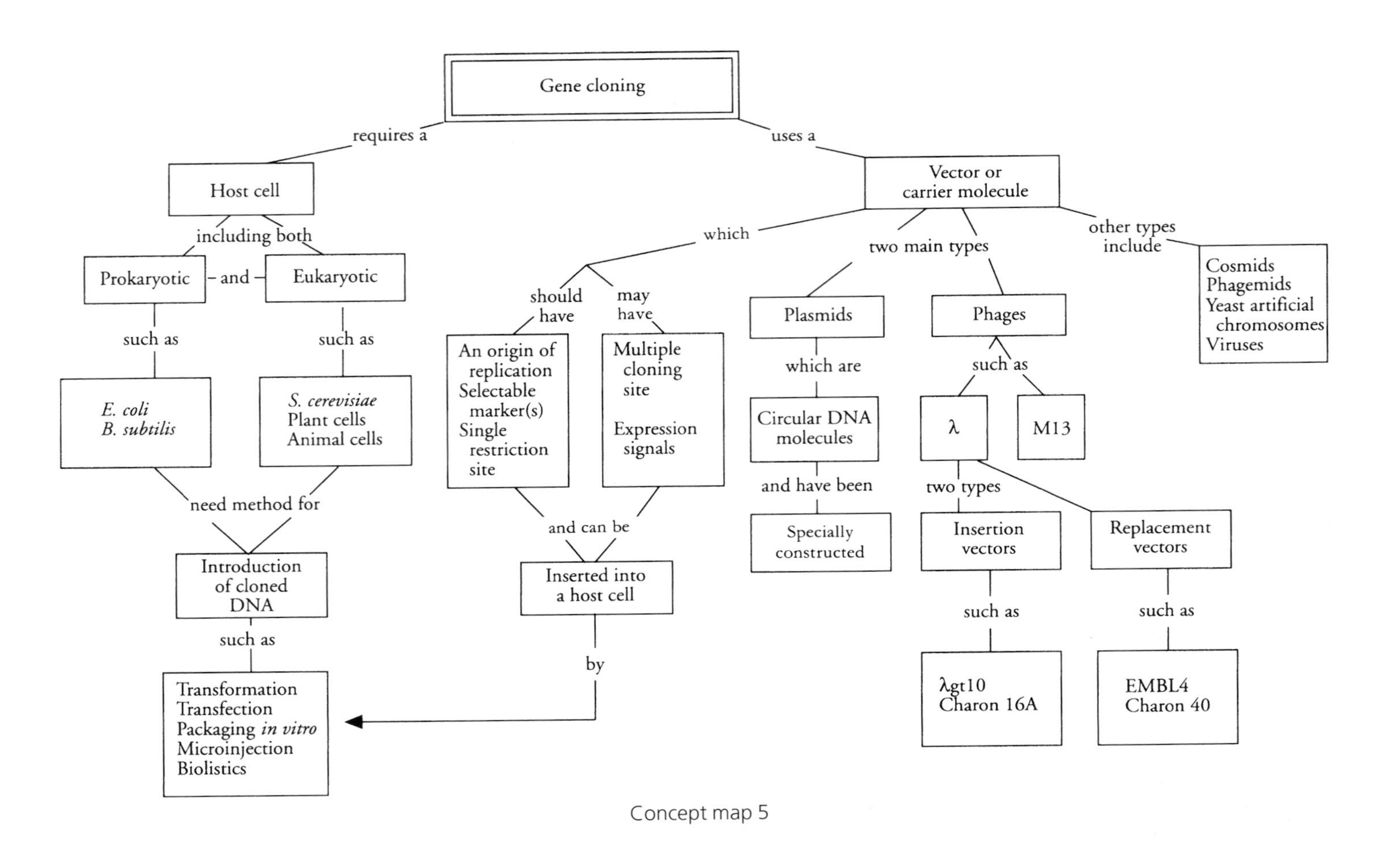

Concept map 5

6

Cloning strategies

In the previous two chapters I examined the two essential components of genetic engineering, these being (i) the ability to cut, modify and join DNA molecules *in vitro*, and (ii) the host/vector systems that allow recombinant DNA molecules to be propagated. With these components at his or her disposal, the genetic engineer has to devise a cloning strategy that will enable efficient use of the technology to achieve the aims of the experiment. In Chapter 1 I showed that there are basically four stages to any cloning experiment (Fig. 1.1), involving **generation** of DNA fragments, **joining** to a vector, **propagation** in a host cell, and **selection** of the required sequence. In this chapter I examine some of the strategies that are available for completing the first three of these stages by the traditional methods of gene cloning, largely restricting the discussion to cloning eukaryotic DNA in *E. coli*. The use of the **polymerase chain reaction** (PCR) in amplification and cloning of sequences is discussed in Chapter 7, as this is now a widely used protocol which in some cases bypasses standard cloning techniques. Selection of cloned sequences is discussed in Chapter 8, although the type of selection method that will be used does have to be considered when choosing host/vector combinations for a particular cloning exercise.

6.1 Which approach is best?

The complexity of any cloning experiment depends largely on two factors: (i) the overall aims of the work, and (ii) the type of source material from which the nucleic acids will be isolated for cloning. Thus a strategy to isolate and

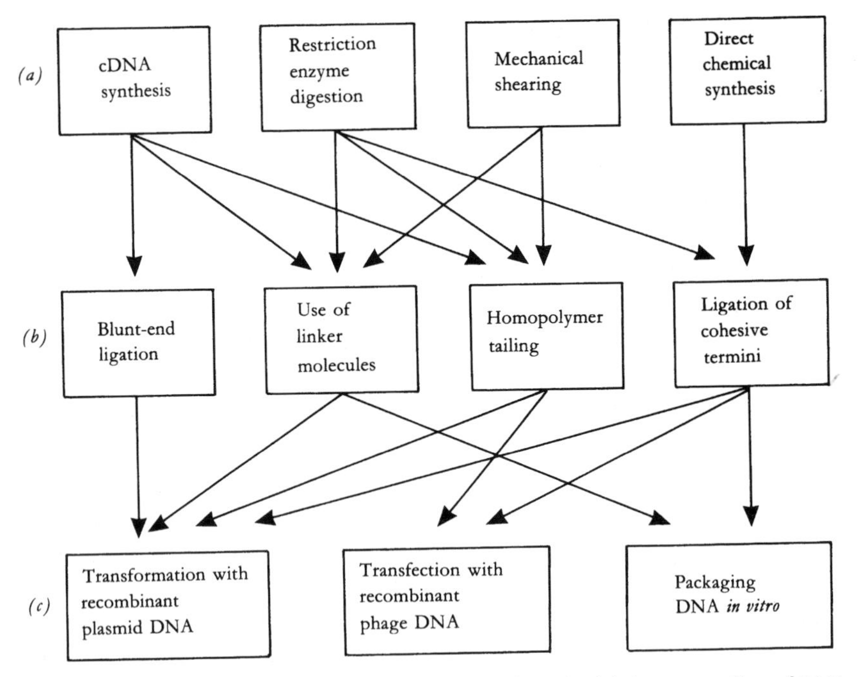

Fig. 6.1. Routes available for cloning. The possibilities for (*a*) the generation of DNA fragments, (*b*) joining to a vector and (*c*) introducing the recombinant DNA into a host cell. Preferred routes are indicated by arrows. Redrawn from Old and Primrose (1989), *Principles of Gene Manipulation*, Blackwell. Reproduced with permission.

sequence a relatively small DNA fragment from *E. coli* will be different (and will probably involve fewer stages) from a strategy to produce a recombinant protein in a transgenic eukaryotic organism. There is no single cloning strategy that will cover all requirements. Each project will therefore be unique, and will present its own set of problems that have to be addressed by choosing the appropriate path through the maze of possibilities (see Fig. 6.1). Fortunately, most of the confusion can be eliminated by careful design of experiments and rigorous interpretation of results.

When dealing with eukaryotic organisms, the first major decision is whether to begin with messenger RNA (mRNA) or genomic DNA. Although the DNA represents the complete genome of the organism, it may contain non-coding DNA such as introns, control regions and repetitive sequences. This can sometimes present problems, particularly if the genome is large and the aim is to isolate a single-copy gene. However, if the primary interest is in the control of gene expression, it is obviously necessary to isolate the control sequences, so genomic DNA is the only alternative.

Messenger RNA has two advantages over genomic DNA as a source material. Firstly, it represents the genetic information that is being expressed by the particular cell type from which it is prepared. This can be a very powerful preliminary selection mechanism, as not all the genomic DNA will be represented in the mRNA population. Also, if the gene of interest is highly expressed, this may be reflected in the abundance of its mRNA, and this can make isolation of the clones easier. A second advantage of mRNA is that it, by definition, represents the coding sequence of the gene, with any introns having been removed during RNA processing. Thus production of recombinant protein is much more straightforward if a clone of the mRNA is available.

Although genomic DNA and mRNA are the two main sources of nucleic acid molecules for cloning, it is possible to synthesise DNA *in vitro* if the amino acid sequence of the protein is known. Whilst this is a laborious task for long stretches of DNA, it is a useful technique in some cases, particularly if only short sections of a gene need to be synthesised to complete a sequence prior to cloning.

Having decided on the source material, the next step is to choose the type of host/vector system. Even when cloning in *E. coli* hosts there is still a wide range of strains available, and care must be taken to ensure that the optimum host/vector combination is chosen. When choosing a vector, the method of joining the DNA fragments to the vector and the means of getting the recombinant molecules into the host cell are two main considerations. In practice the host/vector systems in *E. coli* are well defined, so it is a relatively straightforward task to choose the best combination, given the type of fragments to be cloned and the desired outcome of the experiment. However, the great variety of vectors, host cells and cloning kits available from suppliers can be confusing to the first-time gene manipulator, and often a recommendation from an experienced colleague is the best way to proceed.

In devising a cloning strategy all the points mentioned above have to be considered. Often there will be no ideal solution to a particular problem, and a compromise will have to be accepted. By keeping the overall aim of the experiments in mind, the researcher can minimise the effects of such compromises and choose the most efficient cloning route.

6.2 Cloning from mRNA

Each type of cell in a multicellular organism will produce a range of mRNA molecules. In addition to the expression of general 'housekeeping' genes

Table 6.1. *m*RNA *abundance classes*

Source	Number of different mRNAs	Abundance (molecules/cells)
Mouse liver cytoplasmic poly(A)$^+$ RNA	9	12 000
	700	300
	11 500	15
Chick oviduct polysomal poly(A)$^+$ RNA	1	100 000
	7	4000
	12 500	5

Note: The diversity of mRNAs is indicated by the number of different mRNA molecules. There is one mRNA that is present in chick oviduct cells at a very high level (100 000 molecules per cell). This mRNA encodes ovalbumin, the major egg white protein. Source: After Old & Primrose (1989), *Principles of Gene Manipulation*, 4th edition, Blackwell. Mouse data from Young *et al.* (1976), *Biochemistry* **15**, 2823–2828, copyright (1976) American Chemical Society. Chick data from Axel *et al.* (1976), Cell **11**, 247–254, copyright (1976) Cell Press. Reproduced with permission.

whose products are required for basic cellular metabolism, cells exhibit tissue-specific gene expression. Thus liver cells, kidney cells, skin cells, etc., will each synthesise a different spectrum of tissue-specific proteins (and hence mRNAs). In addition to the **diversity** of mRNAs produced by each cell type, there may well be different **abundance classes** of particular mRNAs. This has important consequences for cloning from mRNA, as it is easier to isolate a specific cloned sequence if it is present as a high proportion of the starting mRNA population. Some examples of mRNA abundance classes are shown in Table 6.1.

6.2.1 Synthesis of cDNA

It is not possible to clone mRNA directly, so it has to be converted into DNA before being inserted into a suitable vector. This is achieved using the enzyme reverse transcriptase (RTase; see Section 4.2.2) to produce **complementary DNA** (also known as **copy DNA** or **cDNA**). The classic early method of cDNA synthesis utilises the poly(A) tract at the 3′ end of the mRNA to bind an oligo(dT) primer, which provides the 3′-OH group required by RTase (Fig. 6.2). Given the four dNTPs and suitable conditions, RTase will synthesise a copy of the mRNA to produce a cDNA·mRNA hybrid. The mRNA can be

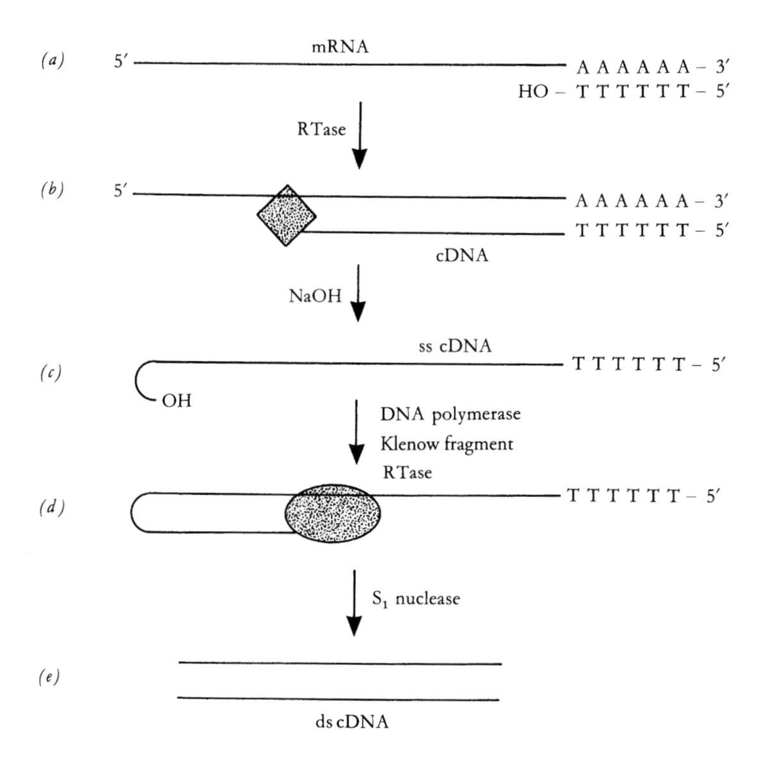

Fig. 6.2. Synthesis of cDNA. Poly(A)$^+$ RNA (mRNA) is used as the starting material. (*a*) A short oligo(dT) primer is annealed to the poly(A) tail on the mRNA, which provides the 3′-OH group for reverse transcriptase to begin copying the mRNA (*b*). The mRNA is removed by alkaline hydrolysis to give a single-stranded cDNA molecule (*c*). This has a short double-stranded hairpin loop structure which provides a 3′-OH terminus for (*d*) second-strand synthesis by a DNA polymerase (T4 DNA polymerase, Klenow fragment, or RTase). (*e*) The double-stranded cDNA is trimmed with S_1 nuclease to produce a blunt-ended ds cDNA molecule. An alternative to the alkaline hydrolysis step is to use RNase H, which creates nicks in the mRNA strand of the mRNA·cDNA hybrid. By using this in conjunction with DNA polymerase I, a nick translation reaction synthesises the second cDNA strand.

removed by alkaline hydrolysis and the single-stranded (ss) cDNA converted into double-stranded (ds) cDNA by using a DNA polymerase. In this second strand synthesis the priming 3′-OH is generated by short hairpin loop regions that form at the end of the ss cDNA. After second strand synthesis, the ds cDNA can be trimmed by S_1 nuclease to give a flush-ended molecule, which can then be cloned in a suitable vector.

Several problems are often encountered in synthesising cDNA using the

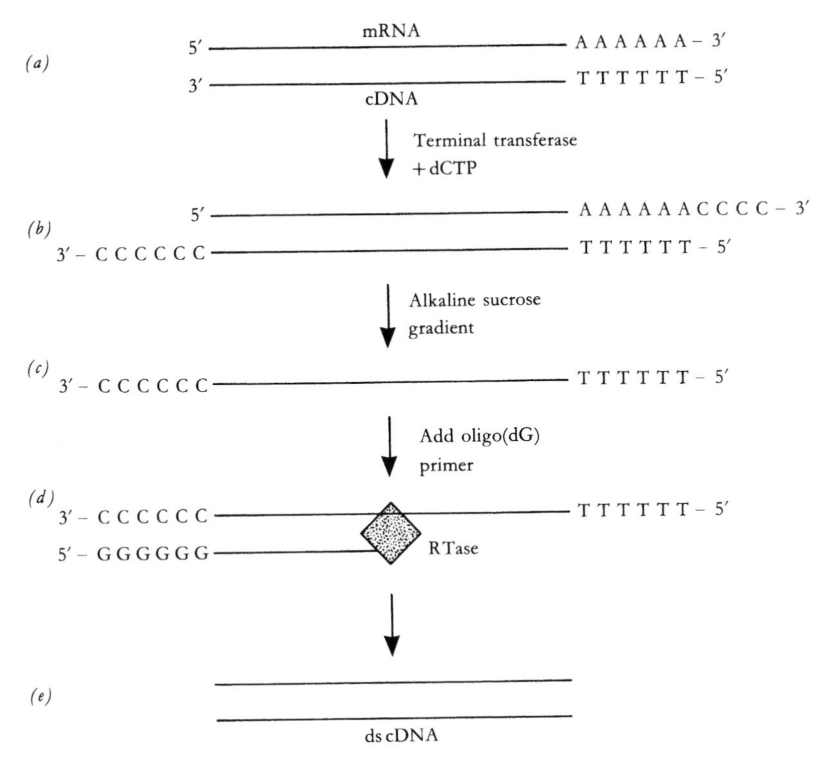

Fig. 6.3. Oligo(dG)-primed second-strand cDNA synthesis. (*a*) First-strand synthesis is as shown in Fig. 6.2, generating an mRNA·cDNA hybrid. (*b*) This is tailed with C residues using terminal transferase. (*c*) Fractionation through an alkaline sucrose gradient hydrolyses the mRNA and permits recovery of full-length cDNA molecules. (*d*) An oligo(dG) primer is annealed to the C tails, and reverse transcriptase used to synthesise the second strand. (*e*) This generates a double-stranded full-length cDNA molecule. From Old and Primrose (1989), *Principles of Gene Manipulation*, Blackwell. (Reproduced with permission.)

method outlined above. Firstly, synthesis of full-length cDNAs may be inefficient, particularly if the mRNA is relatively long. This is a serious problem if expression of the cDNA is required, as it may not contain all the coding sequence of the gene. Such inefficient full-length cDNA synthesis also means that the 3′ regions of the mRNA tend to be over-represented in the cDNA population. Secondly, problems can arise from the use of S_1 nuclease, which may remove some important 5′ sequences when it is used to trim the ds cDNA.

More recent methods for cDNA synthesis overcome the above problems to a great extent, and the original method is now rarely used. One of the simplest adaptations involves the use of oligo(dC) tailing to permit oligo(dG)-primed second-strand cDNA synthesis (Fig. 6.3). The dC tails are added to the 3′ termini of the cDNA using the enzyme terminal transferase. This functions most efficiently on accessible 3′ termini, and the tailing reaction therefore favours full-length cDNAs in which the 3′ terminus is not 'hidden' by the mRNA template. The method also obviates the need for S_1 nuclease treatment, and thus full-length cDNA production is enhanced further.

Many suppliers now produce kits for cDNA synthesis. Often these have been optimised for a particular application, and the number of steps involved is usually reduced to a minimum. In many ways the mystique that surrounded cDNA synthesis in the early days has now gone, and the techniques available make full-length cDNA synthesis a relatively straightforward business. The key to success is to obtain good quality mRNA preparations and to take great care in handling these. In particular, contamination with nucleases must be avoided.

Although the poly(A) tract of eukaryotic mRNAs is often used for priming cDNA synthesis, there may be cases where this is not appropriate. Where the mRNA is not polyadenylated, random oligonucleotide primers may be used to initiate cDNA synthesis. Or, if all or part of the amino acid sequence of the desired protein is known, a specific oligonucleotide primer can be synthesised and used to initiate cDNA synthesis. This can be of great benefit in that specific mRNAs may be copied into cDNA, which simplifies the screening procedure when the clones are obtained. An additional possibility with this approach is to use the polymerase chain reaction (PCR; see Chapter 7) to amplify selectively the desired sequence.

Having generated the cDNA fragments, the cloning procedure can begin. Here there is a further choice to be made regarding the vector system – plasmid or phage, or perhaps cosmid or phagemid? Examples of cloning strategies based on the use of plasmid and phage vectors are given below.

6.2.2 Cloning cDNA in plasmid vectors

Although many workers prefer to clone cDNA using a bacteriophage vector system, plasmids are still often used, particularly where isolation of the desired cDNA sequence involves screening a relatively small number of clones. Joining the cDNA fragments to the vector is usually achieved by one of the three methods outlined in Fig. 6.1 for cDNA cloning, these being **blunt-end ligation**, the use of **linker molecules**, and **homopolymer tailing**. Although favoured for cDNA cloning, these methods may also be used with genomic DNA (see Section 6.3). Each of the three methods will be described briefly.

Blunt-end ligation is exactly what it says – the joining of DNA molecules with blunt ends, using DNA ligase (see Section 4.3). In cDNA cloning, the blunt ends may arise as a consequence of the use of S_1 nuclease, or they may be generated by filling in the protruding ends with DNA polymerase. The main disadvantage of blunt-end ligation is that it is an inefficient process, as there is no specific intermolecular association to hold the DNA strands together whilst DNA ligase generates the phosphodiester linkages required to produce the recombinant DNA. Thus high concentrations of the participating DNAs must be used, so that the chances of two ends coming together are increased. The effective concentration of DNA molecules in cloning reactions is usually expressed as the concentration of termini, thus one talks about 'picomoles of ends', which can seem rather strange terminology to the uninitiated.

The conditions for ligation of ends must be chosen carefully. In theory, when vector DNA and cDNA are mixed, there are several possible outcomes. The desired result is for one cDNA molecule to join with one vector molecule, thus generating a recombinant with one insert. However, if concentrations are not optimal, the insert or vector DNAs may self-ligate to produce circular molecules, or the insert/vector DNAs may form concatemers instead of bimolecular recombinants. In practice, the vector is often treated with a phosphatase (either BAP or CIP; see Section 4.2.3) to prevent self-ligation, and the concentrations of the vector and insert DNAs are chosen to favour the production of recombinants.

One potential disadvantage of blunt-end ligation is that it may not generate restriction enzyme recognition sequences at the cloning site, thus hampering excision of the insert from the recombinant. This is usually not a major problem, as many vectors now have a series of restriction sites clustered around the cloning site. Thus DNA inserted by blunt-end ligation can often be excised by using one of the restriction sites in the cluster. Another approach involves the use of **linkers**, which are self-complementary oligomers that contain a recognition sequence for a particular restriction enzyme. One such sequence would be 5′-CCGAATTCGG-3′, which in double-stranded form will contain the recognition sequence for *Eco*RI (GAATTC). Linkers are synthesised chemically, and can be added to cDNA by blunt-end ligation (Fig. 6.4). When they have been added, the cDNA/linker is cleaved with the linker-specific restriction enzyme, thus generating sticky ends prior to cloning. This can pose problems if the cDNA contains sites for the restriction enzyme used to cleave the linker, but these may be overcome by using a methylase to protect any internal recognition sites from digestion by the enzyme.

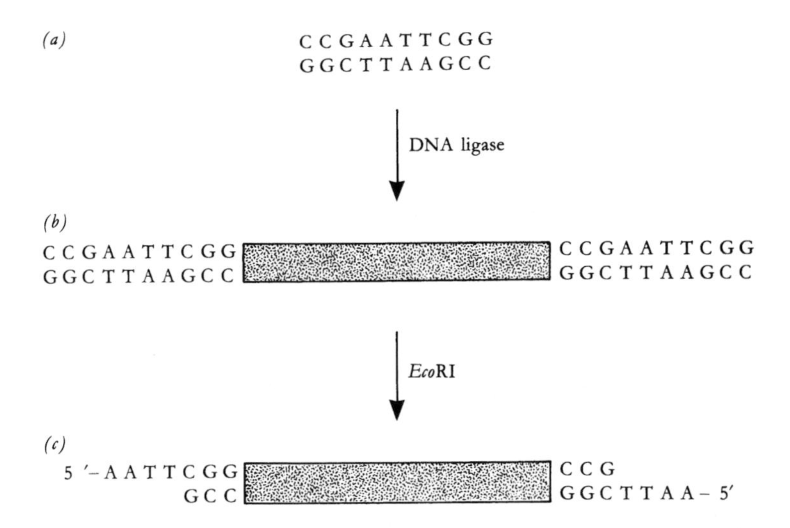

Fig. 6.4. Use of linkers. (*a*) The 10-mer 5′-CCGAATTCGG-3′ contains the recognition site for EcoRI. (*b*) The linker is added to blunt-ended DNA using DNA ligase. (*c*) The construct is then digested with *Eco*RI, which cleaves the linker to generate protruding 5′ termini. Redrawn from Winnacker (1987), *From Genes to Clones*, VCH. (Reproduced with permission.)

A second approach to cloning by addition of sequences to the ends of DNA molecules involves the use of **adaptors** (Fig. 6.5). These are single-stranded non-complementary oligomers that may be used in conjunction with linkers. When annealed together, a linker/adaptor with one blunt end and one sticky end is produced, which can be added to the cDNA to provide sticky-end cloning without digestion of the linkers.

The use of **homopolymer tailing** has proved to be a popular and effective means of cloning cDNA. In this technique, the enzyme terminal transferase (see Section 4.2.3) is used to add homopolymers of dA, dT, dG or dC to a DNA molecule. Early experiments in recombinant production used dA tails on one molecule and dT tails on the other, although the technique is now most often used to clone cDNA into the *Pst*I site of a plasmid vector by dG·dC tailing. Homopolymers have two main advantages over other methods of joining DNAs from different sources. Firstly, they provide longer regions for **annealing** DNAs together than, for example, cohesive termini produced by restriction enzyme digestion. This means that **ligation** need not be carried out *in vitro*, as the cDNA·vector hybrid is stable enough to survive introduction into the host cell, where it is ligated *in vivo*. A second advantage is specificity. As the vector and insert cDNAs have different but complementary 'tails',

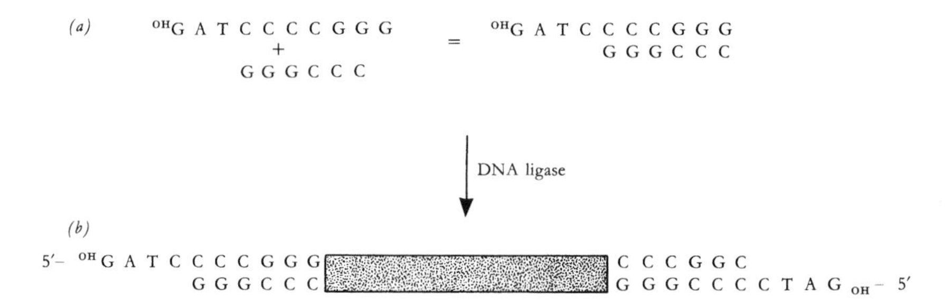

Fig. 6.5. Use of adaptors. In this example a *Bam*HI adaptor (5′-GATCCCCGGG-3′) is annealed with a single-stranded *Hpa*II linker (3′-GGGCCC-5′) to generate a double-stranded sticky-ended molecule, as shown in (*a*). This is added to blunt-ended DNA using DNA ligase. The DNA therefore gains protruding 5′ termini without the need for digestion with a restriction enzyme, as shown in (*b*). The 5′ terminus of the adaptor can be dephosphorylated to prevent self ligation. Redrawn from Winnacker (1987), *From Genes to Clones*, VCH. (Reproduced with permission.)

there is little chance of self-annealing, and the generation of bimolecular recombinants is favoured over a wider range of effective concentrations that is the case for other annealing/ligation reactions.

An example of the use of homopolymer tailing is shown in Fig. 6.6. The vector is cut with *Pst*I and tailed by terminal transferase in the presence of dGTP. This produces dG tails. The insert DNA is tailed with dC in a similar way, and the two can then be annealed. This regenerates the original *Pst*I site, which enables the insert to be cut out of the recombinant using this enzyme.

Introduction of cDNA·plasmid recombinants into suitable *E. coli* hosts is achieved by transformation (Section 5.5.1), and the desired transformants can then be selected by the various methods available (see Chapter 8).

6.2.3 Cloning cDNA in bacteriophage vectors

Although plasmid vectors have been used extensively in cDNA cloning protocols, there are situations where they may not be appropriate. If a large number of recombinants is required, as might be the case if a low-abundance mRNA was to be cloned, phage vectors may be more suitable. The chief advantage here is that packaging *in vitro* may be used to generate the recombinant phage, which greatly increases the efficiency of the cloning process. In addition, it is much easier to store and handle large numbers of phage clones than is the case for bacterial colonies carrying plasmids. Given that isolation of a cDNA

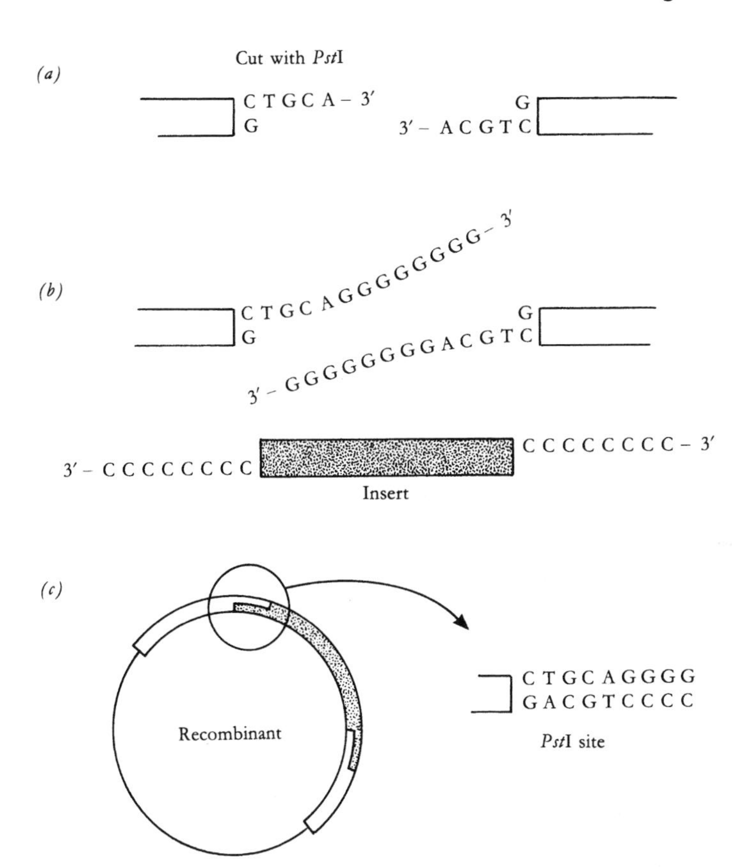

Fig. 6.6. Homopolymer tailing. (*a*) The vector is cut with *PstI*, which generates protruding 3′-OH termini. (*b*) The vector is then tailed with dG residues using terminal transferase. The insert DNA is tailed with dC residues in a similar way. (*c*) The dC and dG tails are complementary and the insert can therefore be annealed with the vector to generate a recombinant. The *Pst*I sites are regenerated at the ends of the insert DNA, as shown.

clone of a rare mRNA species may require screening hundreds of thousands of independent clones, ease of handling becomes a major consideration.

Cloning cDNA in phage λ vectors is, in principle, no different to cloning any other piece of DNA. However, the vector has to be chosen carefully, as cDNA cloning has slightly different requirements than genomic DNA cloning in λ vectors (see Section 6.3). Generally cDNAs will be much shorter than genomic DNA fragments, so an insertion vector is usually chosen. Vectors such as λgt10 and Charon 16A (Section 5.3.2) are suitable, with cloning capacities of some 7.6 and 9.0 kb respectively. The cDNA may be size-fractionated prior to cloning, to remove short cDNAs that may not be representative full-length copies of the

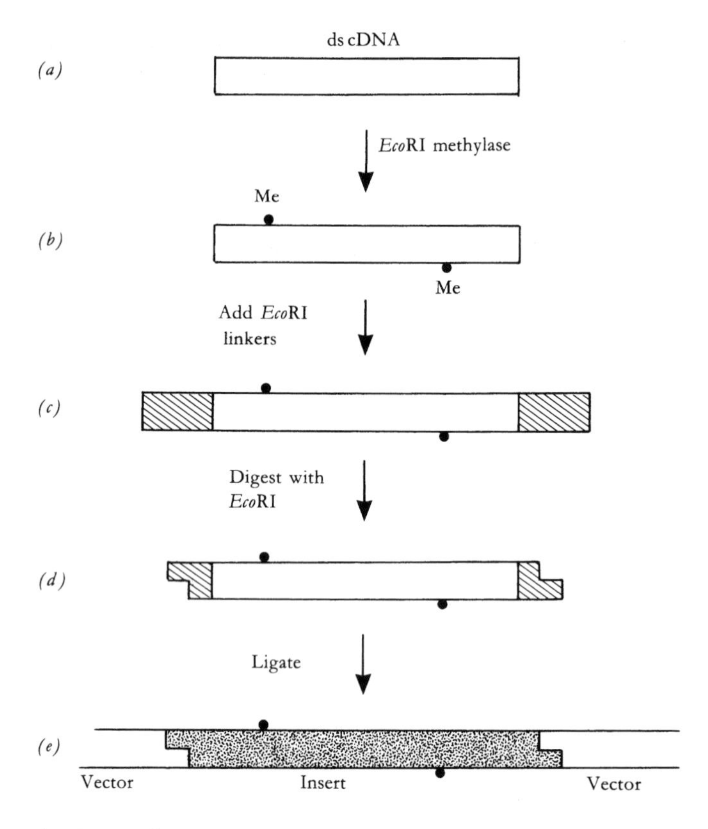

Fig. 6.7. Cloning cDNA in λ vectors using linkers. (*a*) The ds cDNA is treated with *Eco*RI methylase, which (*b*) methylates any internal *Eco*RI recognition sequences. (*c*) *Eco*RI linkers are then added to the ends of the methylated cDNA, and the linkers digested with *Eco*RI. (*d*) The methylation prevents digestion at internal sites, and the result is a cDNA with *Eco*RI cohesive ends. (*e*) This can be ligated into the *Eco*RI site of a λ vector such as λgt10.

mRNA. In the case of vectors such as λgt10, cDNA is usually ligated into the *Eco*RI site using linkers, as shown in Fig. 6.7. The recombinant DNA is packaged *in vitro* and plated on a suitable host for selection and screening.

6.3 Cloning from genomic DNA

Although cDNA cloning is an extremely useful branch of gene manipulation technology, there are certain situations where cDNAs will not provide the answers to the questions that are being posed. If, for example, the overall structure of a particular **gene** is being investigated (as opposed to its RNA transcript), the investigator may wish to determine if there are introns present. He or she will probably also wish to examine the control sequences responsible for regulating gene expression, and these will not be present in the processed mRNA molecule that is represented by a cDNA clone. In such a situation clones generated from genomic DNA must be isolated. This presents a slightly different set of problems than those involved in cloning cDNA, and therefore requires a different cloning strategy.

6.3.1 Genomic libraries

Cloning DNA, by whatever method, gives rise to a population of recombinant DNA molecules, often in plasmid or phage vectors, maintained either in bacterial cells or as phage particles. A collection of independent clones is termed a **clone bank** or **library**. The term **genomic library** is often used to describe a set of clones representing the entire genome of an organism, and the production of such a library is usually the first step in isolating a DNA sequence from an organism's genome.

What are the characteristics of a good genomic library? In theory, a genomic library should represent the entire genome of an organism as a set of overlapping cloned fragments, produced in a random manner, and maintained in a stable form with no misrepresentation of sequences. The systems available for producing genomic libraries essentially fulfil these requirements, although some compromise may be necessary during the cloning process.

The first consideration in constructing a genomic library is the number of clones required. This depends on a variety of factors, the most obvious one being the size of the genome. Thus a small genome such as that of *E. coli* will require many fewer clones than a more complex one such as the human genome. The type of vector to be used also has to be considered, which will determine size of fragments that can be cloned. In practice, library size can be calculated quite simply on the basis of the probability of a particular sequence being represented in the library. There is a formula that takes account of all the factors and produces a 'number of clones' value. The formula is:

$$N = \ln(1 - P)/\ln(1 - a/b)$$

where N is the number of clones required, P is the desired probability of a particular sequence being represented (typically set at 0.95 or 0.99), a is the average size of the DNA fragments to be cloned and b is the size of the genome (expressed in the same units as a).

Table 6.2. *Genomic library sizes for various organisms*

Organism	Genome size (kb)	No. clones N, $P=0.95$	
		20kb inserts	45 kb inserts
Escherichia coli (bacterium)	4.0×10^3	6.0×10^2	2.7×10^2
Saccharomyces cerevisiae (yeast)	1.4×10^4	2.1×10^3	9.3×10^2
Arabidopsis thaliana (simple higher plant)	7.0×10^4	1.1×10^4	4.7×10^3
Drosophila melanogaster (fruit fly)	1.7×10^5	2.5×10^4	1.1×10^4
Stronglyocentrotus purpuratus (sea urchin)	8.6×10^5	1.3×10^5	5.7×10^4
Homo sapiens (human)	3.0×10^6	4.5×10^5	2.0×10^5
Triticum aestivum (hexaploid wheat)	1.7×10^7	2.5×10^6	1.1×10^6

Note: The number of clones (N) required for a probability (P) of 95% that a given sequence is represented is shown for various organisms. The genome sizes of the organisms are given (haploid genome size, if appropriate). Two values of N are shown, for 20kb inserts (λ replacement vector size) and 45 kb inserts (cosmid vectors). The values should be considered as minimum estimates, as strictly speaking the calculation assumes: (i) that the genome size is known accurately, (ii) that the DNA is fragmented in a totally random manner for cloning, (iii) that each recombinant DNA molecule will give rise to a single clone, (iv) that the efficiency of cloning is the same for all fragments, and (v) that diploid organisms are homozygous for all loci. These assumptions are usually not all valid for a given experiment.

By using this formula, it is possible to determine the magnitude of the task ahead, and to plan a cloning strategy accordingly. Some genome sizes and their associated library sizes are shown in Table 6.2. These library sizes should be considered as minimum values, as the generation of cloned fragments may not provide a completely random and representative set of clones in the library. Thus, for a human genomic library, we are talking of some 10^6 clones or more in order to be reasonably sure of isolating a particular single-copy gene sequence.

When dealing with this size of library, phage or cosmid vectors are usually essential, as the cloning capacity and efficiency of these vectors is much greater than that of plasmid vectors. Although cosmids, with the potential to clone fragments of up to 47 kb, would seem to be the better choice, λ replacement vectors are often used for library construction. This is because they are easier to use than cosmid vectors, and this outweighs the disadvantage of having only half the cloning capacity. In addition, the techniques for screening phage libraries are now routine and have been well characterised. This is an important consideration, particularly where workers new to the technology wish to use gene manipulation in their research. Alternatively, artificial chromosome vectors such as BACs or YACs may be used to clone large DNA fragments.

6.3.2 Preparation of DNA fragments for cloning

One of the most important aspects of library production is the generation of genomic DNA fragments for cloning. If a λ replacement vector such as EMBL4 is to be used, the maximum cloning capacity will be around 23 kb. Thus fragments of this size must be available for the production of recombinants. In practice a range of fragment sizes is used, often between 17 and 23 kb for a vector such as EMBL4. It is important that smaller fragments are not ligated into the vector, as there is the possibility of multiple inserts which could arise by ligation of small non-contiguous DNA sequences into the vector.

There are two main considerations when preparing DNA fragments for cloning, these being: (i) the molecular weight of the DNA after isolation from the organism, and (ii) the method used to fragment the DNA. For a completely random library, the starting material should be very high molecular weight DNA, and this should be fragmented by a totally random (i.e. sequence-independent) method. Isolation of DNA in excess of 100 kb in length is desirable, and this in itself can pose technical difficulties where the type of source tissue does not permit gentle disruption of cells. In addition, pipetting and mixing solutions of high molecular weight DNA can cause shearing of the molecules, and great care must be taken when handling the preparations.

Assuming that sufficient DNA of 100 kb is available, fragmentation can be carried out. This is usually followed by a size-selection procedure to isolate fragments in the desired range of sizes. Fragmentation can be achieved either by mechanical shearing or by partial digestion with a restriction enzyme. Although mechanical shearing (by forcing the DNA through a syringe needle,

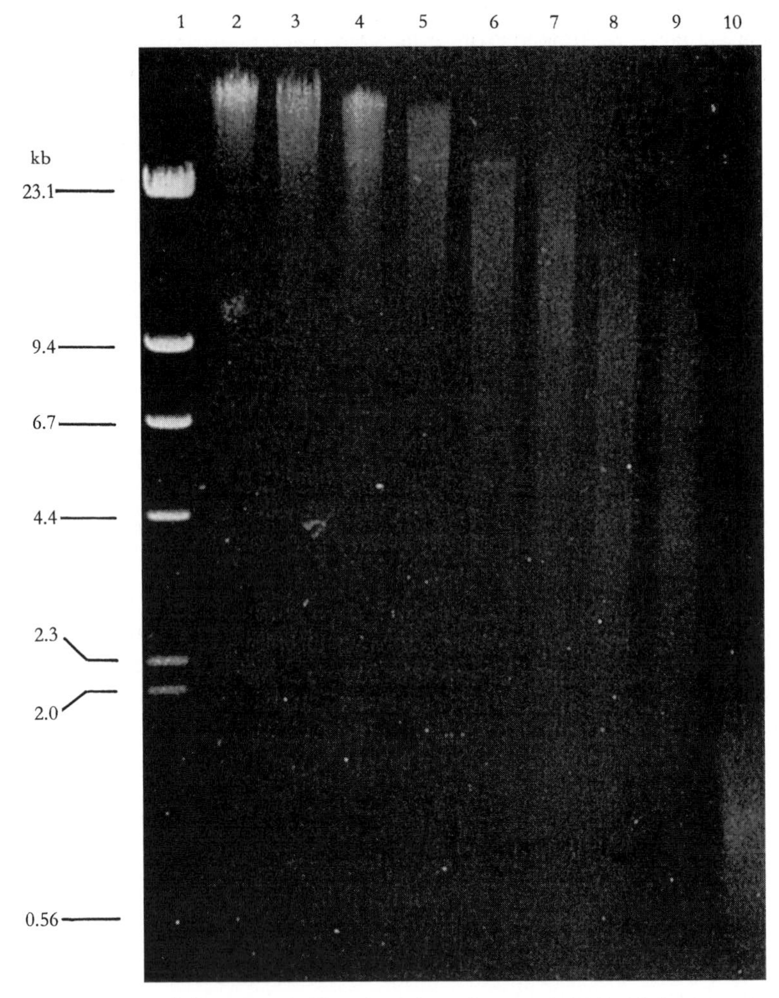

Fig. 6.8. Partial digestion and fractionation of genomic DNA. High molecular weight genomic DNA was digested with various concentrations of the restriction enzyme *Sau3*A. Samples from each digest were run on a 0.7% (w/v) agarose gel and stained with ethidium bromide. Lane 1 shows λ *Hind*III markers, sizes as indicated. Lanes 2 to 10 show the effects of increasing concentrations of restriction enzyme in the digestions. As the concentration of enzyme is increased, the DNA fragments generated are smaller. From this information the optimum concentration of enzyme to produce fragments of a certain size distribution can be determined. These can then be run on a gel (as here) and isolated prior to cloning. (Photograph courtesy of Dr N. Urwin.)

or by sonication) will generate random fragments, it will not produce DNA with cohesive termini. Thus further manipulation such as trimming or filling in the ragged ends of the molecules will be required before the DNA can be joined to the vector, usually with linkers, adaptors or homopolymer tails (see Fig. 6.1). In practice these additional steps are often considered undesirable, and fragmentation by partial restriction digestion is used extensively in library construction. However, this is not a totally sequence-independent process, as the occurrence of restriction enzyme recognition sites is clearly sequence-dependent. Partial digestion is therefore something of a compromise, but careful design and implementation of the procedure can overcome most of the disadvantages.

If a restriction enzyme is used to digest DNA to completion, the fragment pattern will obviously depend on the precise location of recognition sequences. This approach therefore has two drawbacks. Firstly, a six-cutter such as *Eco*RI will have recognition sites on average about once every 4096 bp, which would produce fragments that are too short for λ replacement vectors. Secondly, any sequence bias, perhaps in the form of repetitive sequences, may skew the distribution of recognition sites for a particular enzyme. Thus some areas of the genome may contain few sites, whilst others have an over-abundance. This means that a complete digest will not be suitable for generating a representative library. If, however, a **partial** digest is carried out using an enzyme that cuts frequently (e.g. a four-cutter such as *Sau*3A, which cuts on average once every 256 bp), the effect is to produce a collection of fragments that are essentially random. This can be achieved by varying the enzyme concentration or the time of digestion, and a test run will produce a set of digests which contain different fragment size distribution profiles, as shown in Fig. 6.8.

6.3.3 Ligation, packaging and amplification of libraries

Having established the optimum conditions for partial digestion, a sample of DNA can be prepared for cloning. After digestion the sample is fractionated, either by density gradient centrifugation or by electrophoresis. Fragments in the range 17–23 kb can then be selected for ligation. If *Sau*3A (or *Mbo*I, which has the same recognition sequence) has been used as the digesting enzyme, the fragments can be inserted into the *Bam*HI site of a vector such as EMBL4, as the ends generated by these enzymes are complementary (Fig. 6.9). The insert DNA can be treated with phosphatase to reduce self-ligation or concatemer formation, and the vector can be digested with *Bam*HI and *Sal*I to

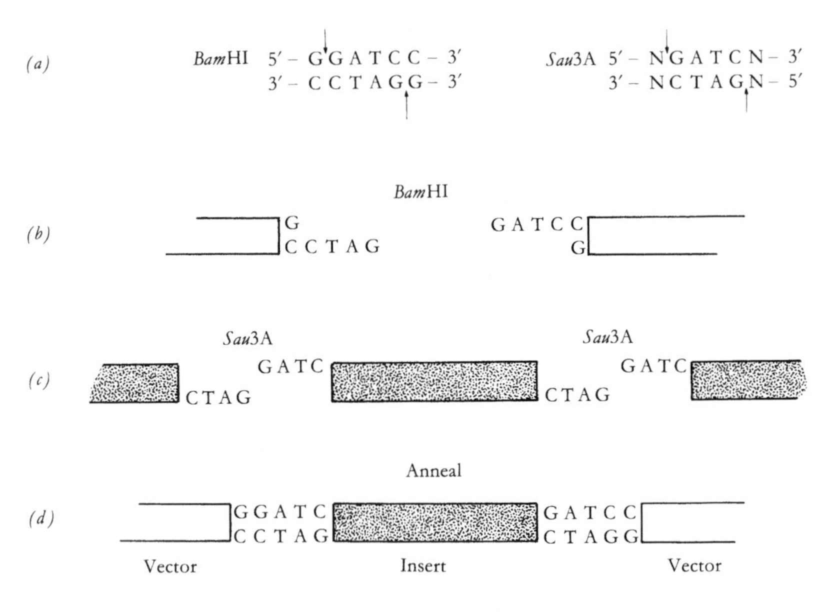

Fig. 6.9. Cloning *Sau*3A fragments into a *Bam*HI site. (*a*) The recognition sequences and cutting sites for *Bam*HI and *Sau*3A. In the *Sau*3A site, N is any base. (*b*) Vector DNA cut with *Bam*HI generates 5′ protruding termini with the four-base sequence 5′-GATC-3′. (*c*) Insert DNA cut with *Sau*3A also generates identical four-base overhangs. (*d*) Thus DNA cut with *Sau*3A can be annealed to *Bam*HI cohesive ends to generate a recombinant DNA molecule.

generate the cohesive ends for cloning and to isolate the stuffer fragment and prevent it from re-annealing during ligation. The *Eco*RI site in the vector can be used to excise the insert after cloning. Ligation of DNA into EMBL4 is summarised in Fig. 6.10.

When ligation is carried out, concatemeric recombinant DNA molecules are produced, which are suitable substrates for packaging *in vitro*, as shown in Fig. 6.11. This produces what is known as a **primary library**, which consists of individual recombinant phage particles. Whilst this is theoretically the most useful type of library in terms of isolation of a specific sequence, it is a finite resource. Thus a primary library is produced, screened and then discarded. If the sequence of interest has not been isolated, more recombinant DNA will have to be produced and packaged. Whilst this may not be a problem, there are occasions where a library may be screened for several different genes, or may be sent to different laboratories. In these cases it is therefore necessary to **amplify** the library. This is achieved by plating the packaged phage on a suitable host strain of *E. coli*, and then resuspending the plaques by gently washing the plates with a buffer solution.

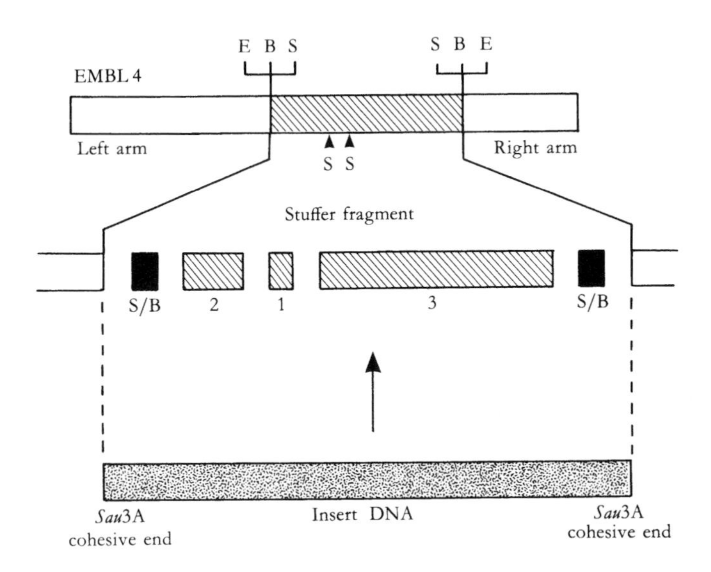

Fig. 6.10. Ligation of *Sau*3A-cut DNA into the λ replacement vector EMBL4. Sites on the vector are *Eco*RI (E), *Bam*HI (B) and *Sal*I (S). The vector is cut with *Bam*HI and *Sal*I, which generates five fragments from the stuffer fragment (hatched in top panel). Removal of the very short *Sal*I/*Bam*HI fragments (filled boxes) prevents the stuffer fragment from re-annealing. In addition, the two internal *Sal*I sites cleave the stuffer fragment, producing three *Sal*I/*Sal*I fragments (1 to 3). If desired, the short fragments can be removed from the preparation by precipitation with isopropanol, which leaves the small fragments in the supernatant. On removal of the stuffer, *Sau*3A-digested insert DNA can be ligated into the *Bam*HI site of the vector (see Fig. 6.9).

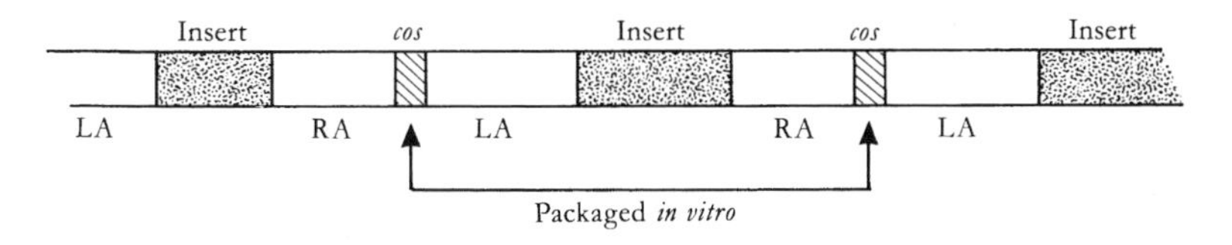

Fig. 6.11. Concatemeric recombinant DNA. On ligation of inserts into a vector such as EMBL4, a concatemer is formed. This consists of the left arm of the vector (LA), the insert DNA, and the right arm (RA). These components of the unit are repeated many times and are linked together at the *cos* sites by the cohesive ends on the vector arms. On packaging *in vitro*, the recombinant genomes are cut at the *cos* sites and packaged into phage heads.

The resulting phage suspension can be stored almost indefinitely, and will provide enough material for many screening and isolation procedures.

Although amplification is a useful step in producing stable libraries, it can lead to skewing of the library. Some recombinant phage may be lost, perhaps due to the presence of repetitive sequences in the insert, which can give rise to recombinational instability. This can be minimised by plating on a recombination-deficient host strain. Some phage may exhibit differential growth characteristics, which may cause particular phage to be either over- or under-represented in the amplified library, and this may mean that a greater number of plaques have to be screened in order to isolate the desired sequence.

6.4 Advanced cloning strategies

In Sections 6.2 and 6.3 I examined cDNA and genomic DNA cloning strategies, using basic plasmid and phage vectors in *E. coli* hosts. These approaches have proved to be both reliable and widely applicable, and still represent a major part of the technology of gene manipulation. However, advances made over the past few years have increased the scope (and often the complexity!) of cloning procedures. Such advances include more sophisticated vectors for *E. coli* and other hosts, increased use of expression vectors, and novel approaches to various technical problems, including the extensive use of PCR technology. Some examples of more advanced cloning strategies are discussed below.

6.4.1 Synthesis and cloning of cDNA

An elegant scheme for generating cDNA clones was developed by Hiroto Okayama and Paul Berg in 1982. In their method the plasmid vector itself is used as the **priming** molecule, and the mRNA is annealed to this for cDNA synthesis. A second **adaptor** molecule is required to complete the process. Both adaptor and primer are based on pBR322, with additional sequences from the SV40 virus. Preparation of the vector and adaptor molecules involves restriction digestion, tailing with oligo(dT) and dG, and purification of the fragments to give the molecules shown in Fig. 6.12. The mRNA is then annealed to the plasmid and the first cDNA strand synthesised and tailed with dC. The terminal vector fragment (which is also tailed during this procedure) is removed and the adaptor added to circularise the vector prior to synthesis

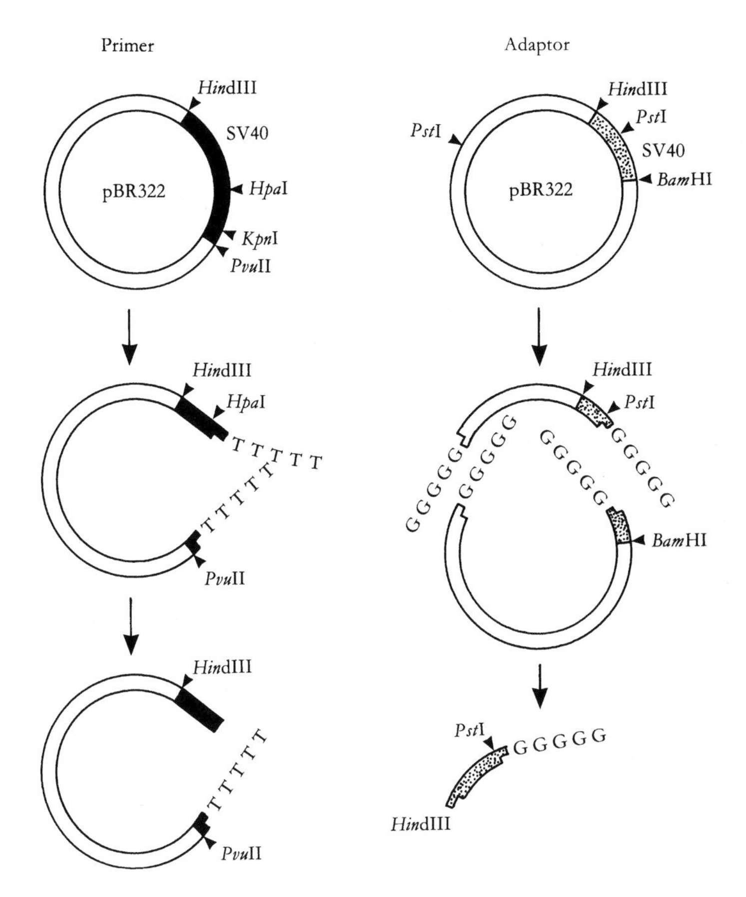

Fig. 6.12. Preparation of vector and adaptor molecules for Okayama and Berg cDNA cloning. The vector is made up from pBR322 plus parts of the SV40 genome (solid or shaded in the diagram). For the primer, the vector is cut with *Kpn*I and tailed with dT residues. It is then digested with *Hpa*I to create a vector in which one end is tailed. The adaptor molecule is generated by cutting the adaptor plasmid with *Pst*I, which generates two fragments. These are tailed with dG residues and digested with *Hind*III to produce the adaptor molecule itself, which therefore has a *Hind*III cohesive end in addition to the dG tail. The fragment is purified for use in the protocol (see Fig. 6.13). From Old and Primrose (1989), *Principles of Gene Manipulation*, Blackwell. (Reproduced with permission.)

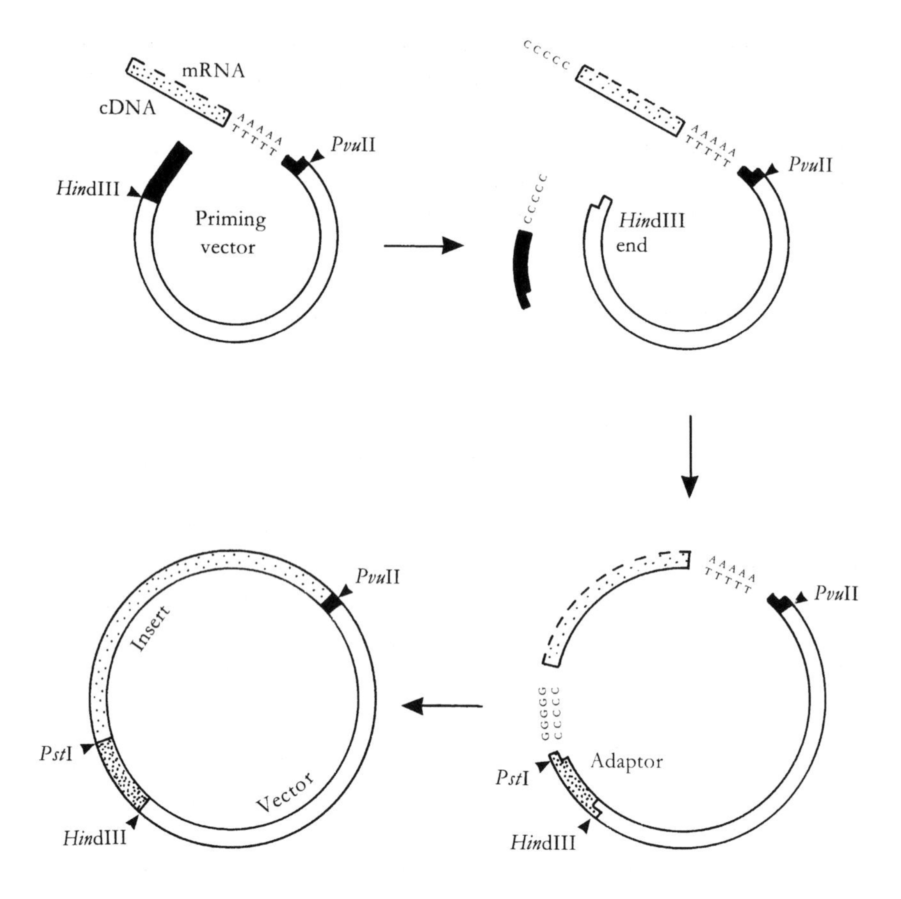

Fig. 6.13. Okayama and Berg cDNA cloning protocol. The mRNA is annealed to the dT-tailed priming vector *via* the poly(A) tail, and the first cDNA strand synthesised using reverse transcriptase. The cDNA is then tailed with dC residues and the dC-tailed vector fragment removed by digestion with *Hind*III. The cDNA is annealed to the dG tailed adaptor molecule, which is ligated into the vector using the cohesive *Hind*III ends on the vector and adaptor molecules. Finally the mRNA is displaced and the second cDNA strand synthesised using RNase H and DNA polymerase I to generate the complete vector/insert recombinant. From Old and Primrose (1989), *Principles of Gene Manipulation*, Blackwell. (Reproduced with permission.)

of the second strand of the cDNA. Second-strand synthesis involves the use of RNase H, DNA polymerase I and DNA ligase in a strand-replacement reaction which converts the mRNA·cDNA hybrid into ds cDNA and completes the ligation of the ds cDNA into the vector. The end result is that recombinants are generated in which there is a high proportion of full-length cDNAs. The Okayama and Berg method is summarised in Fig. 6.13.

6.4.2 Expression of cloned DNA molecules

Many of the routine manipulations in gene cloning experiments do not require expression of the cloned DNA. However, there are certain situations in which some degree of genetic expression is needed. A transcript of the cloned sequence may be required for use as a probe, or a protein product (requiring transcription and translation) may be required as part of the screening process used to identify the cloned gene. Another common biotechnological application is where the recombinant DNA is used to produce a protein of commercial value. If eukaryotic DNA sequences are cloned, post-transcriptional and post-translational modifications may be required, and the type of host/vector system that is used is therefore very important in determining whether or not such sequences will be expressed effectively. The problem of RNA processing in prokaryotic host organisms may be obviated by cloning cDNA sequences, and this is the most common approach where expression of eukaryotic sequences is desired. In this section I consider some aspects of cloning cDNAs for expression, concentrating mainly on the characteristics of the vector/insert combination that enable expression to be achieved. Further discussion of the topic is presented in Chapters 9 and 10.

Assuming that a functional cDNA sequence is available, a suitable host/vector combination must be chosen. The host cell type will usually have been selected by considering aspects such as ease of use, fermentation characteristics or the ability to secrete proteins derived from cloned DNA. However, for a given host cell, there may be several types of expression vector, including both plasmid and (for bacteria) phage-based examples. In addition to the normal requirements such as restriction site availability and genetic selection mechanisms, a key feature of expression vectors is the type of **promoter** that is used to direct expression of the cloned sequence. Often the aim will be to maximise the expression of the cloned sequence, so a vector with a highly efficient promoter is chosen. Such promoters are often termed **strong** promoters. However, if the product of the cloned gene is toxic to the cell, a

Table 6.3. *Promoters used in expression vectors*

Organism	Gene promoter	Induction by
E. coli	*lac* operon	IPTG
	trp operon	β-Indolylacetic acid
	λP_L	Temperature-sensitive λ cI protein
A. nidulans	Glucoamylase	Starch
S. cerevisiae	Acid phosphatase	Phosphate depletion
	Alcohol dehydrogenase	Glucose depletion
	Galactose utilization	Galactose
	Metallothionein	Heavy metals
T. reesei	Cellobiohydrolase	Cellulose
Mouse	Metallothionein	Heavy metals
Human	Heat-shock protein	Temperature >40°C

Note: Some examples of various promoters that can be used in expression vectors are given, with the organism from which the gene promoter is taken. The conditions under which gene expression is induced from such promoters are also given.
Source: Collated from Brown (1990). *Gene Cloning*, Chapman & Hall; and Old & Primrose (1989). *Principles of Gene Manipulation*, Blackwell. Reproduced with permission.

weak promoter may be required to avoid cell death due to over-expression of the toxic product.

Promoters are regions with a specific base sequence, to which RNA polymerase will bind. By examining the base sequence lying on the 5′ (upstream) side of the coding regions of many different genes, the types of sequences that are important have been identified. Although there are variations, these sequences all have some similarities. The 'best fit' sequence for a region such as a promoter is known as the **consensus sequence**. In prokaryotes there are two main regions that are important. Some 10 base-pairs upstream from the transcription start site (the −10 region, as the T_C start site is numbered +1) there is a region known as the **Pribnow box**, which has the consensus sequence 5′-TATAAT-3′. A second important region is located around position −35, and has the consensus sequence 5′-TTGACA-3′. These two regions form the basis of promoter structure in prokaryotic cells, with the precise sequences found in each region determining the strength of the promoter.

Sequences important for transcription initiation in eukaryotes have been identified in much the same way as for prokaryotes. Eukaryotic promoter structure is generally more complex than that found in prokaryotes, and control of initiation of transcription can involve sequences (e.g. enhancers) that may be several hundreds or thousands of base-pairs upstream from the T_C start site. However, there are important motifs closer to the start site. These are a region centred around position −25 with the consensus sequence 5′-TATAAAT-3′ (the **TATA** or **Hogness** box) and a sequence in the −75 region with the consensus 5′-GG(T/C)CAATCT-3′, known as the **CAAT** box.

In addition to the strength of the promoter, it may be desirable to regulate the expression of the cloned cDNA by using promoters from genes that are either **inducible** or **repressible**. Thus some degree of control can be exerted over the transcriptional activity of the promoter; when the cDNA product is required, transcription can be 'switched on' by manipulating the system using an appropriate metabolite. Some examples of promoters used in the construction of expression vectors are given in Table 6.3.

In theory, constructing an expression vector is straightforward once a suitable promoter has been identified. In practice, as is often the case, the process is often highly complex, requiring many manipulations before a functional vector is obtained. The basic vector must carry an origin of replication that is functional in the target host cell, and there may be antibiotic resistance genes or other genetic selection mechanisms present. However, as far as expression of cloned sequences is concerned, it is the arrangement of restriction sites immediately downstream from the promoter that is critical. There must be a unique restriction site for cloning into, and this has to be located in a position where the inserted cDNA sequence can be expressed effectively. This aspect of vector structure is discussed further when the applications of recombinant DNA technology are considered in Part III.

6.4.3 Cloning large DNA fragments in BAC and YAC vectors

Bacterial artificial chromosomes (BACs) and yeast artificial chromosomes (YACs; see Fig. 5.11) can be used to clone very long pieces of DNA. The use of a BAC or YAC vector can reduce dramatically the number of clones needed to produce a representative genomic library for a particular organism, and this is a desirable outcome in itself. A consequence of cloning large pieces of DNA is that physical mapping of genomes is made simpler, as

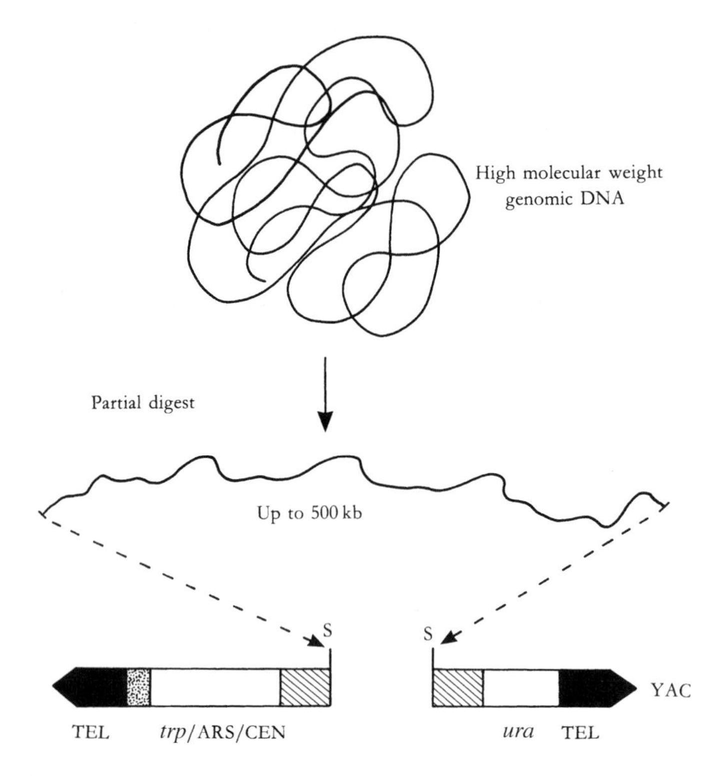

Fig. 6.14. Cloning in a YAC vector. Very large DNA fragments (up to 500 kb) are generated from high molecular weight DNA. The fragments are then ligated into a YAC vector (see Fig. 5.11) that has been cut with *Bam*HI and *Sma*I (S). The construct contains the cloned DNA and the essential requirements for a yeast chromosome, i.e. telomeres (TEL), an autonomous replication sequence (ARS) and a centromere region (CEN). The *trp* and *ura* genes can be used as dual selectable markers to ensure that only complete artificial chromosomes are maintained. From Kingsman and Kingsman (1988), *Genetic Engineering*, Blackwell. (Reproduced with permission.)

there are not as many non-contiguous sequences to fit together in the correct order.

A further advantage of cloning long stretches of DNA stems from the fact that many eukaryotic genes are much larger than the 47 kb or so that can be cloned using cosmid vectors in *E. coli*. Thus with plasmid, phage and cosmid vectors it may be impossible to isolate the entire gene. This makes it difficult to determine gene structure without using several different clones, which is not the ideal way to proceed. The use of BAC or YAC vectors can alleviate this problem and can enable the structure of large genes to be determined by providing a single DNA fragment to work from.

Let's consider using a YAC vector to clone DNA fragments. In practice, cloning in YAC vectors is similar to other protocols (Fig. 6.14). The vector is prepared by a double restriction digest, which releases the vector sequence between the telomeres and cleaves the vector at the cloning site. Thus two arms are produced, as is the case with phage vectors. Insert DNA is prepared as very long fragments (a partial digest with a six-cutter may be used) and ligated into the cloning site to produce artificial chromosomes. Selectable markers on each of the two arms ensure that only correctly constructed chromosomes will be selected and propagated.

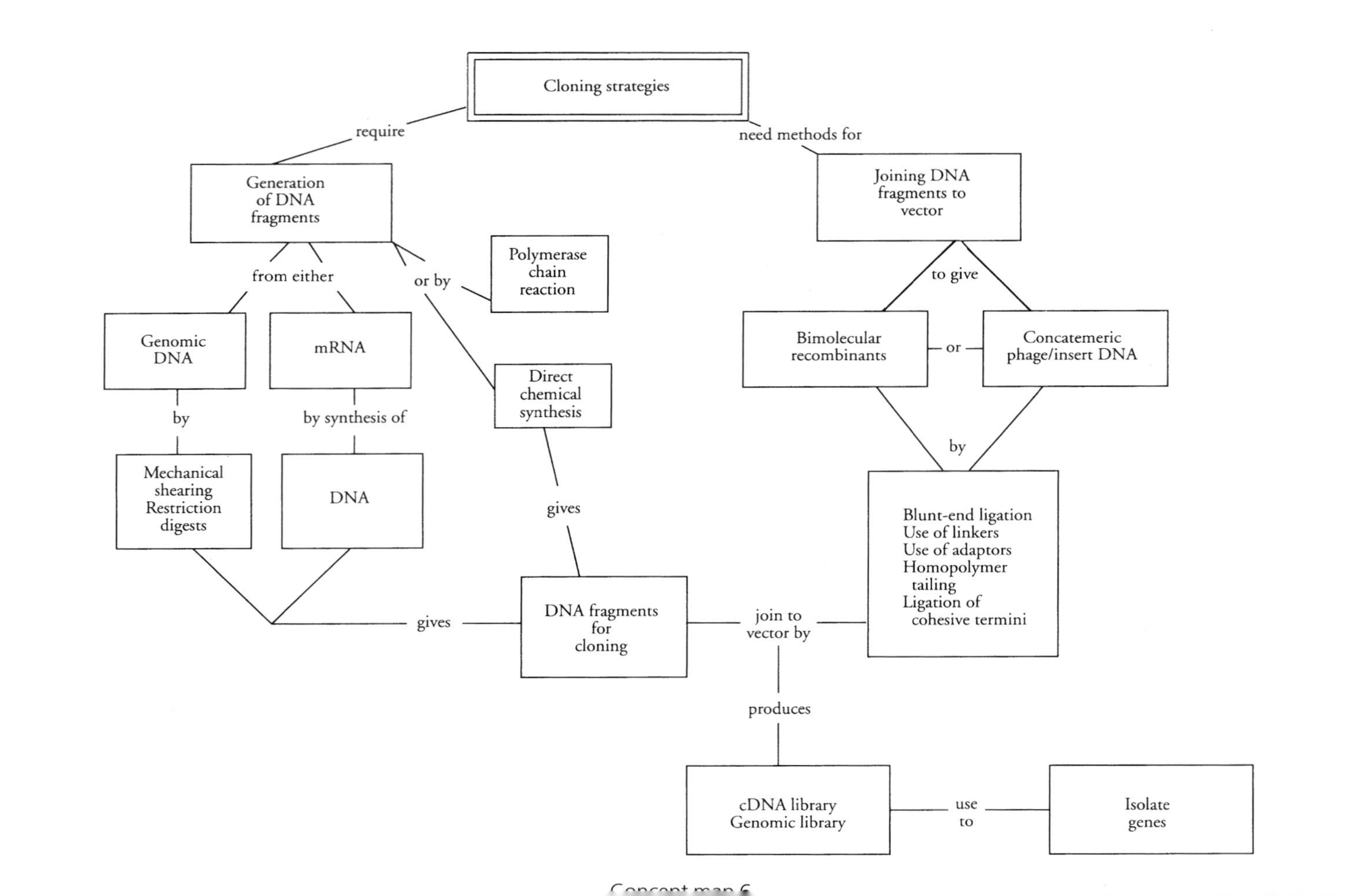

Concept map 6

7

The polymerase chain reaction

Now and again a scientific discovery is made that changes the whole course of the development of a subject. In the field of molecular biology we can identify several major milestones – the emergence of bacterial genetics, the discovery of the mechanism of DNA replication, the double helix and the genetic code, restriction enzymes, and finally the techniques of recombinant DNA. Many of these areas of molecular biology have been recognised by the award of the Nobel prize in either Chemistry or in Medicine and Physiology. Some of these key discoveries are listed in Table 7.1.

The topic of this chapter is the **polymerase chain reaction** (**PCR**), which was discovered by Kary Mullis and for which he was awarded the Nobel prize in Chemistry in 1993. The PCR technique produces a similar result to DNA cloning – the selective amplification of a DNA sequence – and has become such an important part of the genetic engineer's toolkit that in many situations it has essentially replaced traditional cloning methodology. In this chapter we will look at some of the techniques and applications of PCR technology.

7.1 The (short) history of the PCR

The essentials for PCR were in place by the late 1970s. In 1979 Kary Mullis joined the Cetus Corporation, based in Emeryville, California. He was working on oligonucleotide synthesis, which by the early 1980s had become an automated and somewhat tedious process. Thus, his mind was free to investigate other avenues. In his own words, he found himself 'puttering around with oligonucleotides', and the main thrust of his puttering was to try

Table 7.1. *Some milestones in molecular biology recognised by the award of the Nobel prize*

Year	Prize	Recipient(s)	Awarded for studies on
1958	C	Frederick Sanger	Primary structure of proteins
	M&P	Joshua Lederberg	Genetic recombination in bacteria
		George W. Beadle	Gene action
		Edward L. Tatum	
1959	M&P	Arthur Kornberg	Synthesis of DNA and RNA
		Severo Ochoa	
1962	C	John C. Kendrew	3D structure of globular proteins
		Max F. Perutz	
	M&P	Francis H.C. Crick	3D structure of DNA (the double helix)
		James D. Watson	
		Maurice H. F. Wilkins	
1965	M&P	Francois Jacob	Operon theory for bacterial gene expression
		Andre M. Lwoff	
		Jacques L. Monod	
1968	M&P	H. Gobind Khorana	The elucidation of the genetic code and its role in protein synthesis
		Robert W. Holley	
1969	M&P	Max Delbrück	Structure and replication of viruses
		Alfred D. Hershey	
		Salvador E. Luria	
1975	M&P	David Baltimore	Reverse transcriptase and tumour viruses
		Renato Dulbecco	
		Howard M. Temin	
1978	M&P	Werner Arber	Restriction endonucleases
		Daniel Nathans	
		Hamilton O. Smith	
1980	C	Paul Berg	Recombinant DNA technology
		Walter Gilbert	DNA sequencing
		Frederick Sanger	
1982	C	Aaron Klug	Structure of nucleic acid/protein complexes
1984	M&P	George Kohler	Monoclonal antibodies
		Cesar Milstein	
		Niels K. Jerne	Antibody formation
1989	C	Thomas R. Cech	Catalytic RNA
		Sydney Altman	
	M&P	J. Michael Bishop	Genes involved in malignancy
		Harold Varmus	

Table 7.1. (*cont.*)

Year	Prize	Recipient(s)	Awarded for studies on
1993	C	Kary B. Mullis	The polymerase chain reaction
		Michael Smith	Site-directed mutagenesis
	M&P	Richard J. Roberts	Split genes and RNA processing
		Philip A. Sharp	

Note: 'C' and 'M & P' refer to nobel prizes in Chemistry and Medicine & Physiology, respectively. Note also that Frederick Sanger has been awarded *two* Nobel prizes for his work on proteins (1958) and DNA sequencing (1980).

to develop a modified version of the dideoxy sequencing procedure. His thoughts were therefore occupied with oligonucleotides, DNA templates and DNA polymerase.

Late one Friday night in April 1983, Mullis was driving to his cabin with a friend, and was thinking about his modified sequencing experiments. He was, in fact, trying to establish if extension of oligonucleotide primers by DNA polymerase could be used to 'mop up' unwanted dNTPs in the solution, which would otherwise get in the way of his dideoxy experiment. Suddenly he realised that, if *two* primers were involved, and they served to enable extension of the DNA templates, the sequence would effectively be duplicated. Fortunately, he had also been writing computer programs that required reiterative loops – and realised that sequential repetition of his copying reaction (although not what was intended in his experimental system!) could provide many copies of the DNA sequence. Some hasty checking of the figures confirmed that the exponential increase achieved was indeed 2^n, where n is the number of cycles. The PCR had been discovered.

Subsequent work proved that the theory worked when applied to a variety of DNA templates. Mullis presented his work as a poster at the annual Cetus Scientific Meeting in the Spring of 1984. In his account of the discovery of PCR in Scientific American (April 1990), he recalls how Joshua Lederberg discussed his results and appeared to react in a way that was to become familiar – the 'why didn't I think of that' acceptance of a discovery that is brilliant in its simplicity.

Over the past 15 years the PCR technique has been adopted by scientists in a pattern similar to that for recombinant DNA technology itself. The acceptance and use of a procedure can best be demonstrated by looking at the number of published scientific papers in which the technique is used. For PCR, this is shown in Fig. 7.1. This impressive increase in use confirms the

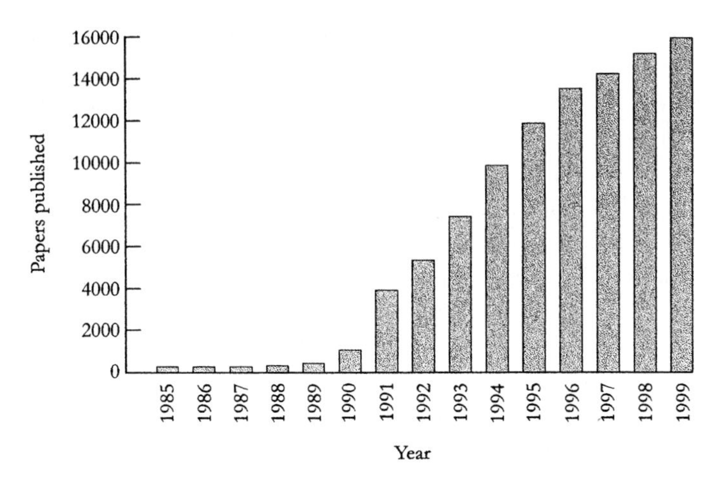

Fig. 7.1. Number of publications citing the polymerase chain reaction from 1985 to 1999. From McPherson and Moller (2000), *PCR*, Bios. (Reproduced with permission.)

importance of PCR, which is now established as one of the major techniques for gene manipulation and analysis.

7.2 The methodology of the PCR

As outlined above, the polymerase chain reaction is elegantly simple in theory. When a DNA duplex is heated, the strands separate or 'melt'. If the single-stranded sequences can be copied by a DNA polymerase, the original DNA sequence is effectively duplicated. If the process is repeated many times, there is an exponential increase in the number of copies of the starting sequence. The length of the fragment is defined by the 5′ ends of the primers, which helps to ensure that a homogeneous population of DNA molecules is produced. Thus, after relatively few cycles, the target sequence becomes greatly amplified, which generates enough of the sequence for identification and further processing.

7.2.1 The essential features of the PCR

In addition to a DNA sequence for amplification, there are two requirements for PCR. Firstly, a suitable primer is required. In practice, **two** primers are necessary, one for each strand of the duplex. The primers should flank the target sequence, so some sequence information is required if selective amplification is to be achieved. The primers are synthesised as oligonucleotides, and are added to the reaction in excess, so that each of the primers is always available following the denaturation step. A second requirement, which makes life much easier for the operator, is the availability of a thermostable form of DNA polymerase. This is purified from the thermophilic bacterium *Thermus aquaticus*, which inhabits hot springs. The use of *Taq* polymerase means that the PCR procedure can be automated, as there is no need to add fresh polymerase after each denaturation step, as would be the case if a heat-sensitive enzyme was used. In addition to these two critical components, the usual mix of the correct buffer composition and the availability of the four dNTPs is needed to ensure that copying of the DNA strands is not stalled due to inactivation of the enzyme or lack of monomers.

In operation, the PCR is straightforward. The target DNA and reaction components are usually mixed together at the start of the process, and the tube heated to around 90 °C to denature the DNA. As the temperature drops, primers will anneal to their target sequences on the single-stranded DNA, and *Taq* polymerase will begin to copy the template strands. The cycle is completed (and re-started) by a further denaturation step. The operational sequence is shown in Fig. 7.2.

Automation of the PCR cycle of operations is achieved by using a programmable heating system known as a **thermal cycler**. This takes small microcentrifuge tubes (96-well plates or glass capillaries can also be used) in which the reactants are placed. Thin-walled tubes permit more rapid temperature changes than standard tubes or plates. Various thermal cycling patterns can be set according to the particular reaction conditions required for a given experiment, but in general the cycle of events shown in Fig. 7.2 forms the basis of the amplification stage of the PCR process. Although thermal cyclers are simple devices, they have to provide accurate control of temperature, and similar rates of heating and cooling for tubes in different parts of the heating block. More sophisticated devices provide a greater range of control patterns than the simpler versions, such as variable rates of heating and cooling, and heated lids to enclose the tubes in a sealed environment. If a heated lid is not used, there is the possibility of evaporation of liquid during the PCR. A layer of mineral or silicone oil on top of the reactants can avoid this, although sometimes contamination of the tube contents with oil can be a problem.

A final practical consideration in setting up a PCR protocol is general housekeeping and manipulation of samples. As the technique is designed to amplify small amounts of DNA, even trace contaminants (perhaps from a tube that has been lying open in a lab, or from an ungloved finger) can sometimes ruin

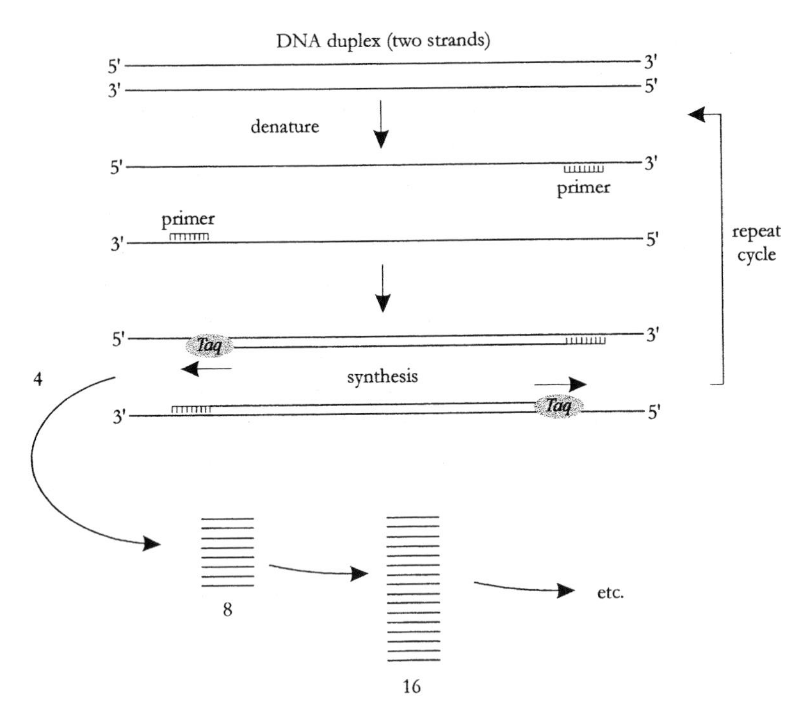

Fig. 7.2. The basic polymerase chain reaction (PCR). Duplex DNA is heat denatured to give single strands, and two oligonucleotide primers are annealed to their complementary sequences on the target DNA. *Taq* polymerase (thermostable) is used to synthesise complementary strands from the template strands by primer extension. The cycle is then repeated by denaturation of the DNA, and the denature/prime/copy programme repeated many times. A typical temperature profile would be denaturation at 95 °C, primer annealing at 55 °C, and extension by *Taq* polymerase at 72 °C. The numbers refer to the number of DNA strands in the reaction; there are four at the end of the first cycle, eight at the end of the second, 16 at the end of the third, and so on. In theory, after 30 cycles 5.4×10^9 strands are produced. Thus DNA sequences can be amplified very quickly.

an experiment. Thus the operator needs to be fastidious (or even a little paranoid!) about cleanliness when carrying out PCR. Also, even the aerosols created by pipetting reagents can lead to cross-contamination, so good technique is essential. It is best if a sterile hood or flow cabinet can be set aside for setting up the PCR reactions, with a separate area used for post-reaction processing. Accurate labelling of tubes (primers, target DNAs, nucleotide mixes, etc.) is also needed, and quality control procedures are important, particularly where analysis is being carried out in medical or forensic applications.

7.2.2 The design of primers for PCR

Oligonucleotide primers are available from many commercial sources, and can be synthesised to order in a few days. In designing primers, there are several aspects that have to be considered. Perhaps most obvious is the **sequence** of the primer – more specifically, where does the sequence information come from? It may be derived from amino acid sequence data, in which case the degeneracy of the genetic code has to be considered, as shown in Fig. 7.3. In synthesising the primer, two approaches can be taken. By incorporating a mixture of bases at the wobble position, a mixed probe can be made, with the 'correct' sequence represented as a small proportion of the mixture. Alternatively, the base **inosine** (which pairs equally well with any of the other bases) can be incorporated as the third base in degenerate codons.

If the primer sequence is taken from an already determined DNA sequence, this may be from the same gene from a different organism, or may be from a cloned DNA that has been sequenced during previous experimental work.

Regardless of the source of the sequence information for the primers, there are some general considerations that should be addressed. The **length** of the primer is important. It should be long enough to ensure stable hybridisation to the target sequence at the required temperature. Although calculation of the melting temperature (T_m) can be used to provide information about annealing temperatures, this is often best determined empirically. The primer must also be long enough to ensure that it is a **unique sequence** in the genome from which the target DNA is taken. Primer lengths of around 20–30 nucleotides are usually sufficient for most applications. With regard to the base composition and sequence of primers, repetitive sequences should be avoided, and also regions of single-base sequence. Primers should obviously not contain regions of internal complementary sequence, or regions of sequence overlap with other primers.

As extension of PCR products occurs from the 3′ termini of the primer, it is this region that is critical with respect to fidelity and stability of pairing with the target sequence. Some 'looseness' of primer design can be accommodated at the 5′ end, and this can sometimes be used to incorporate design features such as restriction sites at the 5′ end of the primer.

7.2.3 DNA polymerases for PCR

Originally the Klenow fragment of DNA polymerase was used for PCR, but this is thermolabile and requires addition of fresh enzyme for each extension *(a)* **Selecting the sequence**

amino acid sequence	Phe -	Leu -	Pro -	Ser -	Ala -	Lys -	Trp -	Ala -	Tyr -	Asp -	Pro
number of codons per amino acid	2	(6)	4	(6)	4	2	1	4	2	2	4
			avoid				better sequence				

(b) **Mixed probe synthesis**

```
Ala Lys Trp Ala Tyr Asp Pro

GCAAAATGGGCATACGACCC
  G  G     G  T  T
  C        C
  T        T
```

number of possibilities 4 x 2 x 1 x 4 x 2 x 2 x 1 = 128

(c) **Using inosine as the degenerate base**

```
Ala Lys Trp Ala Tyr Asp Pro

GCIAAATGGGCITACGACCC
     G        T  T
```

number of possibilities 1 x 2 x 1 x 1 x 2 x 2 x 1 = 8

Fig. 7.3. Primer design. In (*a*) the amino acid sequence and number of codons per amino acid are shown. Those amino acids with six codons (circled) are best avoided. The boxed sequence is therefore selected. In (*b*) a mixed probe is synthesised by including the appropriate mixture of dNTPs for each degenerate position. Note that in this example the final degenerate position for proline is not included, giving an oligonucleotide with 20 bases (a 20-mer). There are 128 possible permutations of sequence in the mixture. In (*c*) inosine (**I**) is used to replace the fourfold degenerate bases, giving eight possible sequences.

phase of the cycle. This was inefficient in that the operator had to be present at the machine for the duration of the process, and a lot of enzyme was needed. Also, as extension was carried out at 37 °C, primers could bind to non-target regions, generating a high background of non-specific amplified products. The availability of *Taq* polymerase solved these problems. Today, a wide variety of thermostable polymerases is available for PCR, sold under licence from Hoffman LaRoche (who hold the patent rights for the use of *Taq* polymerase in PCR). These include several versions of recombinant *Taq* polymerase, as well as enzymes from *Thermus flavus* (*Tfl* polymerase) and *Thermus thermophilus* (*Tth* polymerase).

The key features required for a DNA polymerase include **processivity** (affinity for the template, which determines the number of bases incorporated before dissociation), **fidelity** of incorporation, **rate of synthesis** and the **half-life** of the enzyme at different temperatures. In theory the variations in these aspects shown by different enzymes should make choice of a polymerase a difficult one; in reality, a particular source is chosen and conditions adjusted empirically to optimise the activity of the enzyme.

Of the features mentioned above, fidelity of incorporation of nucleotides is perhaps the most critical. Obviously an error-prone enzyme will generate mutated versions of the target sequence out of proportion to the basal rate of mis-incorporation, given the repetitive cycling nature of the reaction. In theory, an error rate of 1 in 10^4 in a millionfold amplification would produce mutant sequences in around a third of the products. Thus steps often need to be taken to identify and avoid such mutated sequences in cases where high fidelity copying is essential.

7.3 More exotic PCR techniques

As the PCR technique became established, variations of the basic procedure were developed. This is still an area of active development, and new techniques and applications for PCR appear regularly. In this section we will look at some of the variations on the basic PCR process.

7.3.1 PCR using mRNA templates

This is a commonly used variant of the basic PCR, known as **reverse transcriptase PCR** (**RT-PCR**). It can be useful in determining low levels of gene expression by analysing the PCR product of a cDNA prepared from the mRNA transcript. In theory a single mRNA molecule can be amplified, but this is unlikely to be achieved (or required) in practice. The process involves copying the mRNA using reverse transcriptase, as in a standard cDNA synthesis. Oligo(dT)-primed synthesis is often used to generate the first strand cDNA. PCR primers are then used as normal, although the first few cycles may be biased in favour of copying the cDNA single-stranded product until enough copies of the second strand have been generated to allow exponential amplification from both primers. This has no

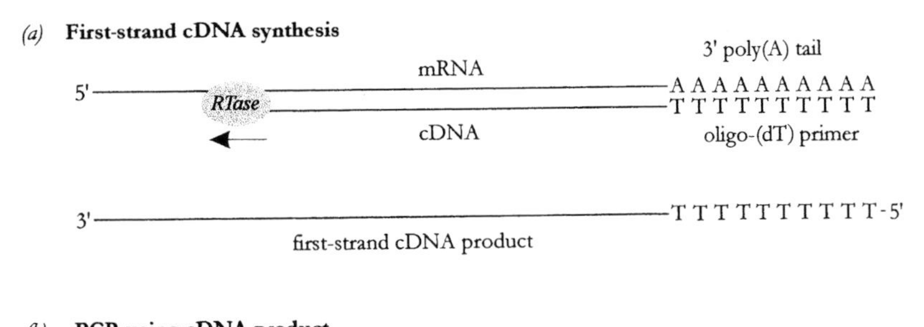

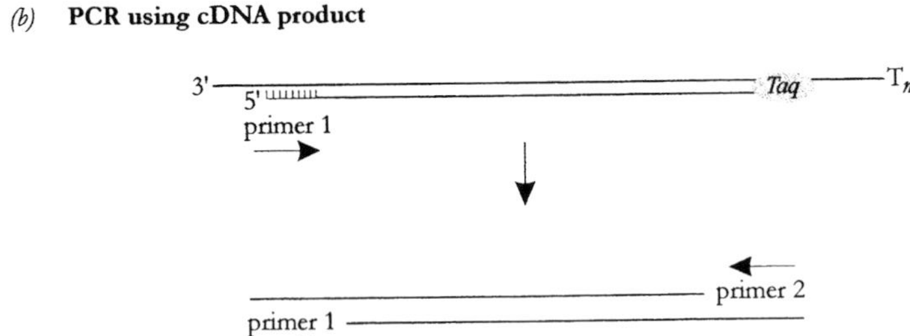

Fig. 7.4. RT-PCR. In (*a*) reverse transcriptase is used to synthesise a cDNA copy of the mRNA. In this example oligo(dT)-primed synthesis is shown. In (*b*) the cDNA product is amplified using gene-specific primers. The initial PCR synthesis will copy the cDNA to give a duplex molecule, which is then amplified in the usual way. In many kits available for RT-PCR the entire procedure can be carried out in a single tube.

effect on the final outcome of the process. An overview of RT-PCR is shown in Fig. 7.4.

One use of RT-PCR is in determining the amount of mRNA in a sample (**competitor RT-PCR**). A differing but known amount of competitor RNA is added to a series of reactions, and the target and competitor amplified using the same primer pair. If the target and competitor products are of different sizes, they can be separated on a gel and the amount of target estimated by comparing with the amount of competitor product. When the two bands are of equal intensity, the amount of target sequence in the original sample is the same as the amount of competitor added. This approach is shown in Fig. 7.5.

7.3.2 Nested PCR

Nested PCR is a useful way of overcoming some of the problems associated with a large number of PCR cycles, which can lead to error-prone synthesis.

(*a*) **Spike RNA samples and convert to cDNA**

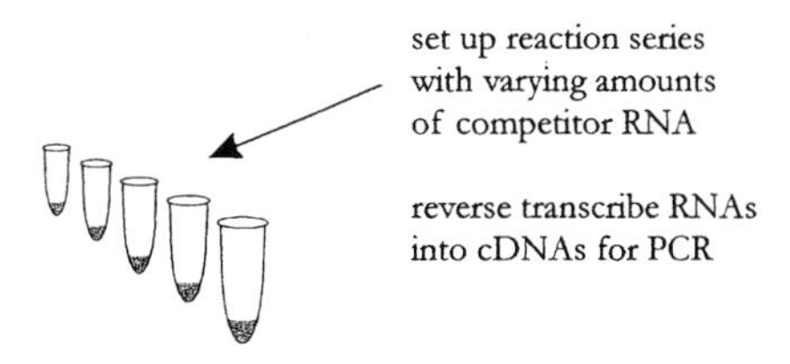

(*b*) **Perform PCR using same primer pair**

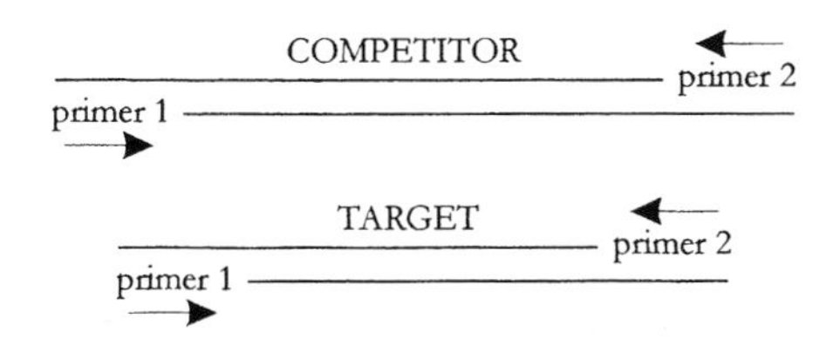

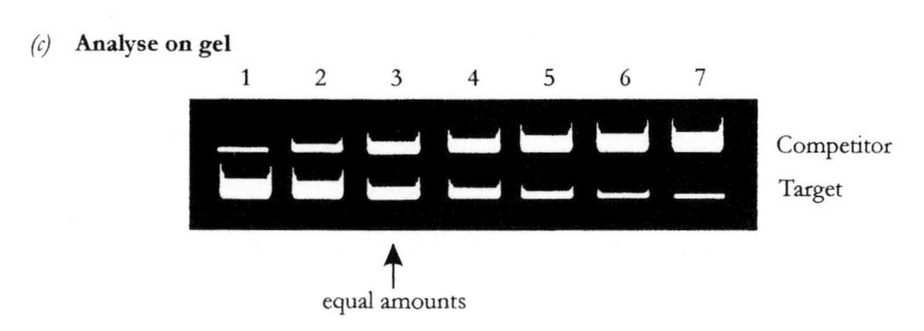

Fig. 7.5. Competitor RT-PCR. (*a*) A series of samples is spiked with increasing amounts of a competitor RNA that will produce a PCR product that is different in size to the prospective target fragment. (*b*) The competitor RNA 'competes' with the target RNA for the primers and other resources in the reaction during PCR. (*c*) When analysed on an electrophoresis gel, the products can be distinguished on the basis of size. In this case the equal band intensities in lane 3 (boxed) enable the amount of target mRNA in the original sample to be determined, as this is the point where the competitor and target were present at equal concentrations at the start of the PCR process.

The technique essentially increases both the sensitivity and fidelity of the basic PCR protocol. It involves using two sets of primers. The first **external** set generates a normal PCR product. Primers that lie inside the first set are then used for a second PCR reaction. These **internal** or **nested** primers generate a shorter product, as shown in Fig. 7.6.

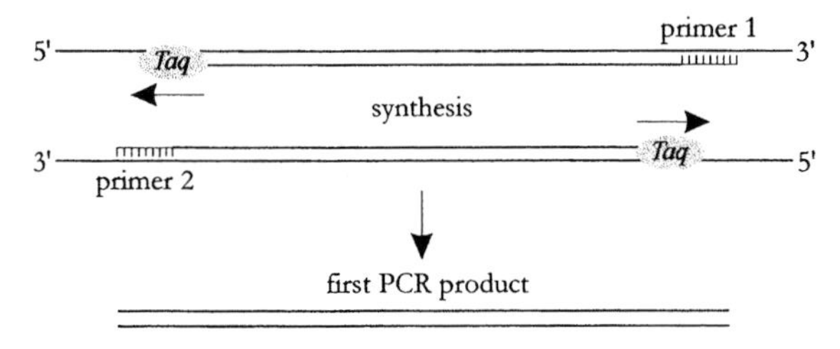

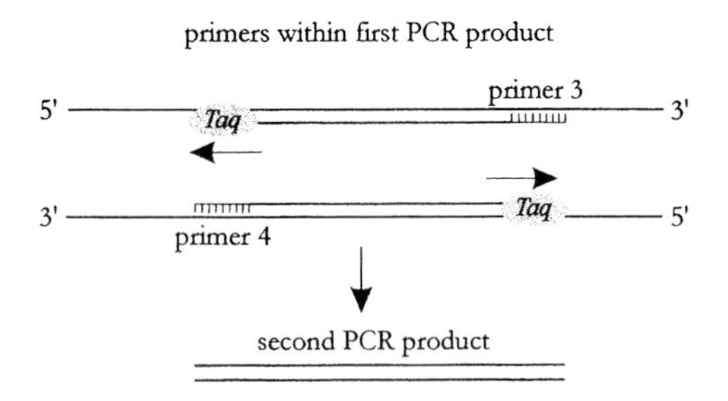

Fig. 7.6. Nested PCR. (*a*) Standard PCR amplification of a fragment using a primer pair. (*b*) In the second PCR, a set of primers that lie inside the first pair is used to prime synthesis of a shorter fragment. Nested PCR can be used to increase the specificity and fidelity of the PCR procedure.

7.3.3 Inverse PCR

Often, a stretch of DNA sequence is known, but the desired target sequence lies outside this region. This causes problems with primer design, as there may be no way of determining a suitable primer sequence for the unknown region. **Inverse PCR (IPCR)** involves isolating a restriction fragment that contains the known sequence plus flanking sequences. By circularising the fragment, and then cutting inside the known sequence, the fragment is essentially inverted. Primers can then be synthesised using the known sequence data and used to amplify the fragment, which will contain the flanking regions. Primers that face away from each other (with respect to direction of product synthesis) in the original known sequence are required, so that on circularisation they are in the correct orientation. The technique can also be used with sets of primers for nested PCR. Deciphering the result usually requires DNA sequencing to determine the areas of interest. Inverse PCR is shown in Fig. 7.7.

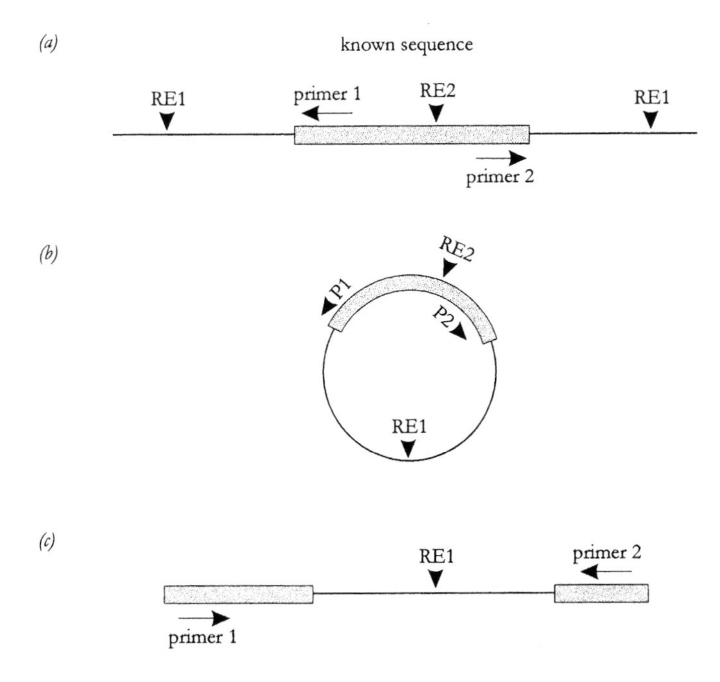

Fig. 7.7. Inverse PCR. (*a*) A region of DNA in which part of the sequence is known (shaded). If the areas of interest lie outside the known region, inverse PCR can be used to amplify these flanking regions. Primers are selected to bind within the known sequence in the opposite orientation to that normally required. Restriction sites are noted as RE1 and RE2. If RE1 is used to cut the DNA, the resulting fragment can be circularised to give the construct shown in (*b*). If this is cut with RE2, the linear fragment that results has the unknown sequence in the middle of two known sequence fragments. Note that the unknown region is non-contiguous, being made up of two flanking regions joined at RE1. The PCR product will therefore contain these two regions plus the regions of known sequence.

7.3.4 RAPD and several other acronyms

Normally the aim of PCR is to generate defined fragments from highly specific primers. However, there are some techniques based on low-stringency annealing of primers. Most widely used of these is **RAPD-PCR**. This stands for **random amplification of polymorphic DNA**. The technique is also known as **arbitrarily primed PCR** (**AP-PCR**). It is a useful method of genomic fingerprinting, and involves using a single primer at low stringency. The primer binds to many sites in the genome, and fragments are amplified from these. The stringency of primer binding can be increased after a few

Table 7.2. *Some variants of the basic PCR process*

Acronym	Technique name	Application
RT-PCR	Reverse transcriptase PCR	PCR from mRNA templates
IPCR	Inverse PCR	PCR of sequences lying outside primer binding sites
RAPD-PCR	Random amplified polymorphic DNA	Genomic fingerprinting under low stringency conditions
AP-PCR	Arbitrarily primed PCR	As RAPD–PCR
TAIL-PCR	Thermal interlaced asymmetric PCR	PCR using alternating high/low primer binding stringency with arbitrary and sequence-specific primers
mrPCR	Multiplex restriction site PCR	PCR using primers with restriction site recognition sequences at their 3′ ends
AFLP	Amplified fragment length polymorphism	Genome analysis by PCR of restriction digests of genomic DNA
CAPS	Cleaved amplified polymorphic sequence analysis	Genome analysis for single nucleotide polymorphisms (SNPs) by PCR and restriction enzyme digestion
GAWTS	Gene amplification with transcript sequencing	PCR coupled with transcript synthesis and sequencing using reverse transcriptase
RAWTS	RNA amplification with transcript sequencing	Variant of GAWTS using mRNA templates
RACE	Rapid amplification of cDNA ends	Isolation of cDNAs from low abundance mRNAs

cycles, which effectively means that only the 'best mismatches' are fully amplified. By careful design the protocol can yield genome-specific band patterns that can be used for comparative analysis. However, there can be problems associated with the reproducibility of the technique, as it is difficult to obtain similar levels of primer binding in different experiments. This is largely due to the mismatched binding of primers at low stringency that is the basis of the technique.

There are many other variants of the PCR protocol, which have given rise (as we have seen) to many new PCR acronyms. Some of these are listed in Table 7.2. These variants will not be discussed further in this book; the reader is directed to the texts in the Section for further reading for further information on the increasingly novel and complex procedures and applications associated with PCR.

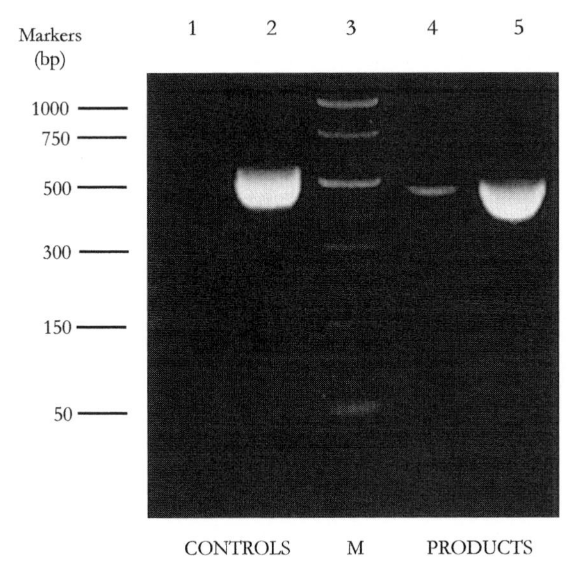

Fig. 7.8. Visualisation of PCR products of ornithine decarboxylase on an agarose gel. Lane 1 – negative control (no DNA); lane 2 – positive control (cloned ornithine decarboxylase fragment, 460 bp); lane 3 – PCR size markers; lanes 4 and 5 – PCR product using rat liver genomic DNA and the ornithine decarboxylase primers used in lane 2. Lane 4 shown product after 15 cycles, lane 5 after 30 cycles of PCR. Photograph courtesy of Dr F. McKenzie.

7.4 Processing of PCR products

Once the PCR process has been completed, the DNA fragments that have been amplified can be analysed. The fate of PCR products depends on the experiment – often a simple visualisation of the product on an electrophoresis gel, with confirmation of the length of the sequence, will be sufficient. This may be coupled with blotting and hybridization techniques to identify specific regions of the sequence. Perhaps the PCR product is to be cloned into an expression vector, in which case specific vectors can be used. Often DNA sequencing is carried out on the PCR product to ensure that the 'correct' sequence has been cloned, which is particularly important when the potential error rate is considered (see Section 7.2.3).

Gel electrophoresis of PCR products is usually the first step in many post-reaction processes. This can verify the fragment size, and can also give some idea of its purity and homogeneity. A typical PCR result is shown in Fig. 7.8.

7.5 Applications of the PCR

Applications of PCR technology are many and diverse. It can be used to clone specific sequences, although in many cases it is, in fact, not necessary to do this, as enough material for subsequent manipulations may be produced by the PCR process itself. It can be used to clone genes from one organism by using priming sequences from another, if some sequence data are available for the gene in question. Another use of the PCR process is in forensic and diagnostic procedures such as the examination of body fluid stains, or in antenatal screening for genetic disorders. These areas are particularly important in the wider context of the applications of gene technology, and will be discussed further in Part III of this book.

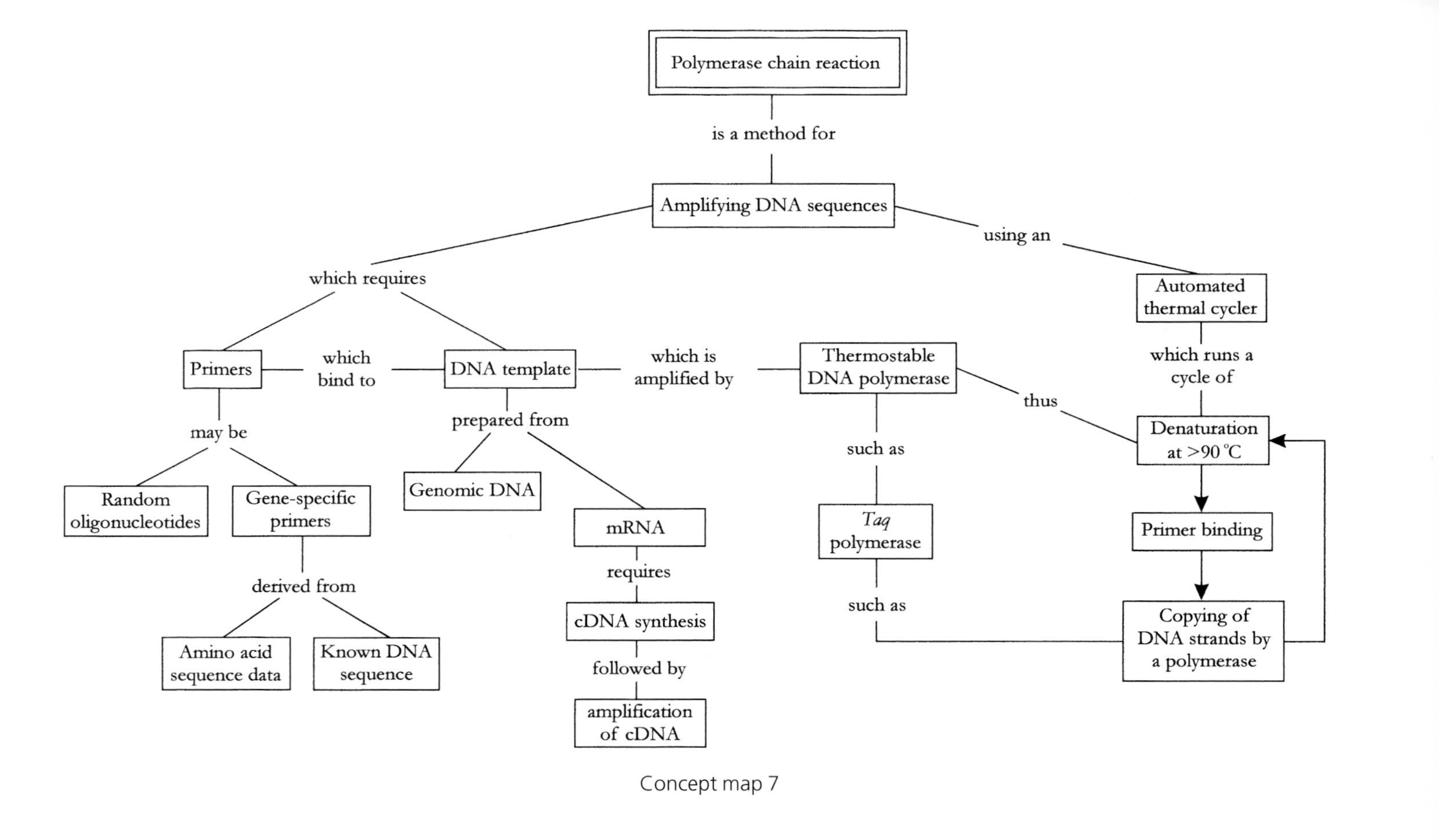

Concept map 7

8

Selection, screening and analysis of recombinants

In this final chapter of Part II, the various techniques that can be used to identify cloned genes will be described. As with previous chapters, the basis of techniques that are perhaps not so widely used today will be included, to illustrate the principles of gene identification and characterisation. This will lead into the final section of the book, where various applications of the technology will be covered, and where we get a look at some of the more advanced developments in gene manipulation.

Success in any cloning experiment depends on being able to identify the desired gene sequence among the many different recombinants that may be produced. Given that a large genomic library may contain a million or more cloned sequences, which are not readily distinguishable from each other by simple analytical methods, it is clear that identification of the target gene is potentially the most difficult part of the cloning process. Fortunately there are several selection/identification methods that can be used to overcome most of the problems that arise.

There are two terms that require definition before we proceed, these being **selection** and **screening**. Selection is where some sort of pressure (e.g. the presence of an antibiotic) is applied during the growth of host cells containing recombinant DNA. The cells with the desired characteristics are therefore **selected** by their ability to survive. This approach ranges in sophistication, from simple selection for the presence of a vector, up to direct selection of cloned genes by complementation of defined mutations. **Screening** is a procedure by which a population of viable cells is subjected to some sort of analysis that enables the desired sequences to be identified. Because only a small proportion of the large number of bacterial colonies or bacteriophage plaques being screened will contain the DNA sequence(s) of interest, screening requires methods that are highly sensitive and specific. In practice, both selection and screening methods may be required in any single experiment, and may even be used at the same time if the procedure is designed carefully.

8.1 Genetic selection and screening methods

Genetic selection and screening methods rely on the expression (or non-expression) of certain traits. Usually, these traits are encoded by the vector, or perhaps by the desired cloned sequence if a direct selection method is available.

One of the simplest genetic selection methods involves the use of antibiotics to select for the presence of vector molecules. For example, the plasmid pBR322 contains genes for ampicillin resistance (Ap^r) and tetracycline resistance (Tc^r). Thus the presence of the plasmid in cells can be detected by plating potential transformants on an agar medium that contains either (or both) of these antibiotics. Only cells that have taken up the plasmid will be resistant, and these cells will therefore grow in the presence of the antibiotic. The technique can also be used to identify mammalian cells containing vectors with selectable markers.

Genetic selection methods can be simple (as above) or complex, depending on the characteristics of the vector/insert combination, and on the type of host strain used. Such methods are extremely powerful, and there is a wide variety of genetic selection and screening techniques available for many diverse applications. Some of these are described below.

8.1.1 The use of chromogenic substrates

The use of chromogenic substrates in genetic screening methods has been an important aspect of the development of the technology. The most popular system uses the compound X-gal (5-bromo-4-chloro-3-indolyl-β-D-galactopyranoside), which is a colourless substrate for β-galactosidase. The enzyme is normally synthesised by *E. coli* cells when lactose becomes available. However, induction can also occur if a lactose analogue such as IPTG (*iso*-propyl-thiogalactoside) is used. This has the advantage of being an inducer without being a substrate for β-galactosidase. On cleavage of X-gal a blue-coloured product is formed (Fig. 8.1), thus the expression of the *lacZ* (β-galactosidase) gene can be detected easily. This can be used either as a

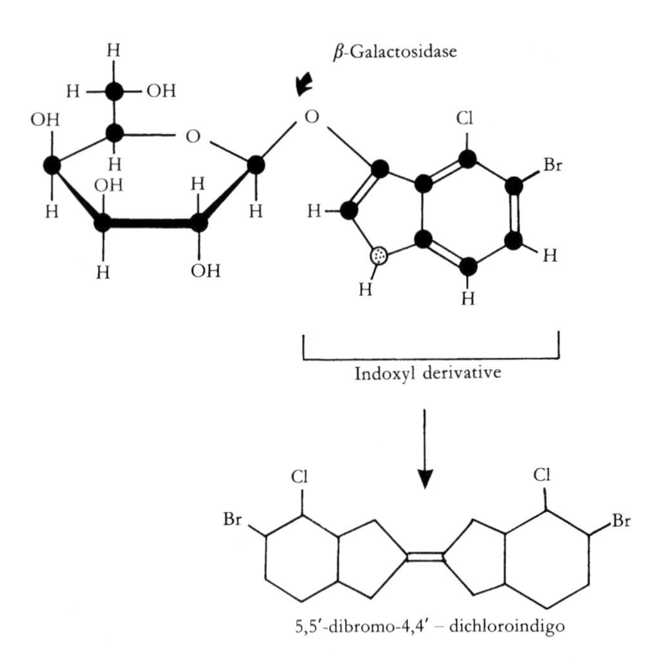

Fig. 8.1. Structure of X-gal and cleavage by β-galactosidase. The colourless compound X-gal (5-bromo-4-chloro-3-indolyl-β-D-galactopyranoside) is cleaved by β-galactosidase to give galactose and an indoxyl derivative. This derivative is in turn oxidised in air to generate the dibromo–dichloro derivative, which is blue.

screening method for cells or plaques, or as a system for the detection of tissue-specific gene expression in transgenics.

The X-gal detection system can be used where a functional β-galactosidase gene is present in the host/vector system. This can occur in two ways. Firstly, an intact β-galactosidase gene (*lacZ*) may be present in the vector, as is the case for the λ insertion vector Charon 16A (see Fig. 5.8). Host cells that are Lac$^-$ are used for propagation of the phage, so that the Lac$^+$ phenotype will only arise when the vector is present. A second approach is to employ the α-complementation system, in which part of the *lacZ* gene is carried by the vector and the remaining part is carried by the host cell. The smaller, vector-encoded peptide fragment is known as the α-peptide, and the region coding for this is designated *lacZ′*. Host cells are therefore designated *lacZ′*$^-$. Blue colonies or plaques will only be produced when the host and vector fragments complement each other to produce functional β-galactosidase.

8.1.2 Insertional inactivation

The presence of cloned DNA fragments can be detected if the insert interrupts the coding sequence of a gene. This approach is known as **insertional inactivation**, and can be used with any suitable genetic system. Three systems will be described to illustrate the use of the technique.

Antibiotic resistance can be used as an insertional inactivation system if DNA fragments are cloned into a restriction site within an antibiotic-resistance gene. For example, cloning DNA into the *Pst*I site of pBR322 (which lies within the Ap^r gene) interrupts the coding sequence of the gene, and renders it non-functional. Thus cells that harbour a recombinant plasmid will be Ap^sTc^r. This can be used to identify recombinants as follows: if transformants are plated firstly onto a tetracycline-containing medium, all cells that contain the plasmid will survive and form colonies. If a replica of the plate is then taken and grown on ampicillin-containing medium, the recombinants (Ap^sTc^r) will not grow, but any non-recombinant transformants (Ap^rTc^r) will. Thus recombinants are identified by their absence from the replica plate, and can be picked from the original plate and used for further analysis.

The X-gal system can also be used as a screen for cloned sequences. If a DNA fragment is cloned into a functional β-galactosidase gene (e.g. into the *Eco*RI site of Charon 16A), any recombinants will be genotypically *lacZ*$^-$ and will therefore not produce β-galactosidase in the presence of IPTG and X-gal. Plaques containing such phage will therefore remain colourless. Non-recombinant phage will retain a functional *lacZ* gene, and therefore give rise to blue plaques. This approach can also be used with the α-complementation system; in this case the insert DNA inactivates the *lacZ'* region in vectors such as the M13 phage and pUC plasmid series. Thus complementation will not occur in recombinants, which will be phenotypically Lac$^-$ and will therefore give rise to colourless plaques or colonies (Fig. 8.2).

Plaque morphology can also be used as a screening method for certain λ vectors such as λgt10, which contain the *cI* gene. This gene encodes the cI repressor, which is responsible for the formation of lysogens. Plaques derived from *cI*$^+$ vectors will be slightly turbid, due to the survival of some cells that have become lysogens. If the *cI* gene is inactivated by cloning a fragment into a restriction site within the gene, the plaques are clear and can be distinguished from the turbid non-recombinants. This system can also be used as a selection method (see Section 8.1.4).

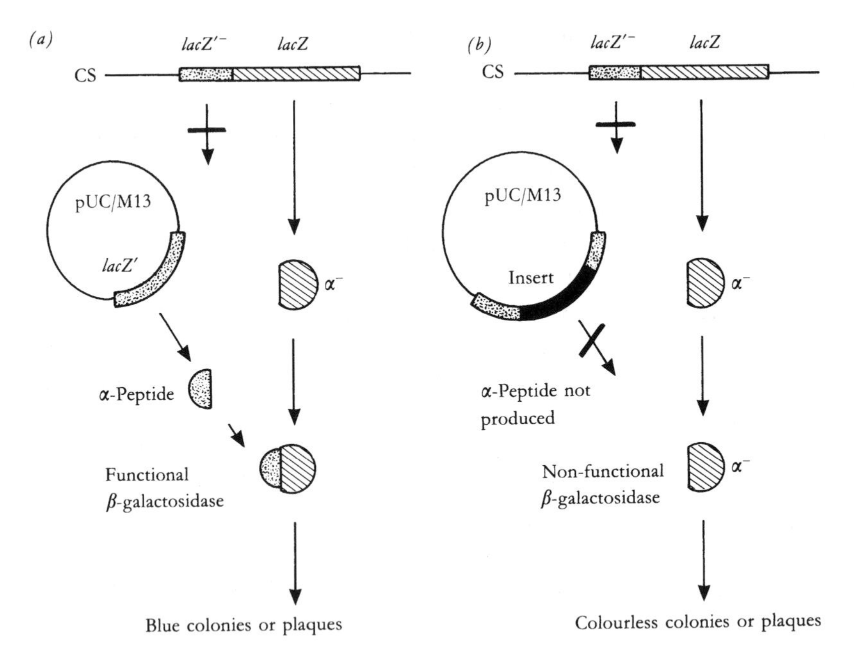

Fig. 8.2. Insertional inactivation in the α-complementation system. (*a*) the chromosome (CS) has a defective *lacZ* gene that does not encode the N-terminal α-peptide of β-galactosidase (specified by the *lacZ*′⁻ gene fragment). Thus the product of the chromosomal *lacZ* region is an enzyme lacking the α-peptide (α^-, hatched). If a non-recombinant pUC plasmid or M13 phage is present in the cell, the *lacZ*′ gene fragment encodes the α-peptide, which enables functional β-galactosidase to be produced. In the presence of X-gal, blue colonies or plaques will appear. If a DNA fragment is cloned into the vector, as shown in (*b*), the *lacZ*′ gene is inactivated and no complementation occurs. Thus colonies or plaques will not appear blue.

8.1.3 Complementation of defined mutations

Direct selection of cloned sequences is possible in some cases. An example is where antibiotic resistance genes are being cloned, as the presence of cloned sequences can be detected by plating cells on a medium that contains the antibiotic in question (assuming that the host strain is normally sensitive to the antibiotic). The method is also useful where specific mutant cells are available, as the technique of **complementation** can be employed, where the cloned DNA provides the function that is absent from the mutant. There are three requirements for this approach to be successful. Firstly, a mutant strain that is deficient in the particular gene that is being sought must be available. Secondly, a suitable selection medium is required, on which the specific recombinants will grow. The final requirement, which is often the limiting step as far as this method is concerned, is that the gene sequence must be expressed in the host cell to give a functional product that will complement the mutation. This is not a problem if, for example, *E. coli* is used to select cloned *E. coli* genes, as the cloned sequences will obviously function in the host cells. This approach has been used most often to select genes that specify nutritional requirements, such as enzymes of the various biosynthetic pathways. Thus genes of the tryptophan operon can be selected by plating recombinants on mutant cells that lack specific functions in this pathway (**auxotrophic** mutants). In some cases, complementation in *E. coli* can be used to select genes from other organisms such as yeast, if the enzymes are similar in terms of their function and they are expressed in the host cell. Complementation can also be used if mutants are available for other host cells, as is the case for yeast and other fungi.

Selection processes can also be used with higher eukaryotic cells. The gene for mouse dihydrofolate reductase (DHFR) has been cloned by selection in *E. coli* using the drug trimethoprim in the selection medium. Cells containing the mouse DHFR gene were resistant to the drug, and were therefore selected on this basis.

8.1.4 Other genetic selection methods

Although the methods outlined above represent some of the ways by which genetic selection and screening can be used to detect the presence of recombinants, there are many other examples of such techniques. These are often dependent on the use of a particular vector/host combination, which enables exploitation of the genetic characteristics of the system. Two examples will be used to illustrate this approach; many others can be found in some of the texts listed in Suggestions for further reading.

The use of the cI repressor system of λgt10 can be extended to provide a powerful **selection** system if the vector is plated on a mutant strain of *E. coli* that produces lysogens at a high frequency. Such strains are designated *hfl* (***h****igh* ***f****requency of* ***l****ysogeny*), and any phage that encodes a functional cI repressor will form lysogens on these hosts. These lysogens will be immune to further infection by phage. DNA fragments are inserted into the λgt10 vector at a restriction site in the *cI* gene. This inactivates the gene, and thus only recombinants (genotypically cI^-) will form plaques.

A second example of genetic selection based on phage/host characteristics is the Spi selection system that can be used with vectors such as EMBL4. Wild-type λ will not grow on cells that already carry a phage, such as phage P2, in the lysogenic state. Thus the λ phage is said to be Spi^+ (***s***ensitive to ***P***2 ***i***nhibition). The Spi^+ phenotype is dependent on the *red* and *gam* genes of λ, and these are arranged so that they are present on the stuffer fragment of EMBL4. Thus recombinants, which lack the stuffer fragment, will be red^- gam^- and will therefore be phenotypically Spi^-. Such recombinants will form plaques on P2 lysogens, whereas non-recombinant phage that are red^+ gam^+ will retain the Spi^+ phenotype and will not form plaques.

8.2 Screening using nucleic acid hybridisation

General aspects of nucleic acid hybridisation are described in Section 3.4. It is a very powerful method of screening clone banks, and is one of the key techniques in gene manipulation. The production of a cDNA or genomic DNA library is often termed the 'shotgun' approach, as a large number of essentially random recombinants is generated. By using a defined nucleic acid **probe**, such libraries can be screened and the clone(s) of interest identified. The conditions for hybridisation are now well established, and the only limitation to the method is the availability of a suitable probe.

8.2.1 Nucleic acid probes

The power of nucleic acid hybridisation lies in the fact that complementary sequences will bind to each other with a very high degree of fidelity (see Fig. 2.9). In practice this depends on the degree of **homology** between the hybridising sequences, and usually the aim is to use a probe that has been derived from the same source as the target DNA. However, under certain conditions sequences that are not 100% homologous can be used to screen for a particular gene, as may be the case if a probe from one organism is used to detect clones prepared using DNA from a second organism. Such **heterologous** probes have been extremely useful in identifying many genes from different sources.

There are three main types of DNA probe, these being (i) cDNA, (ii) genomic DNA and (iii) oligonucleotides. Alternatively, RNA probes can be used if these are suitable. The availability of a particular probe will depend on what is known about the target gene sequence. If a cDNA clone has already been obtained and identified, the cDNA can be used to screen a genomic library and isolate the gene sequence itself. Alternatively, cDNA may be made from mRNA populations and used without cloning the cDNAs. This is often used in what is known as the 'plus/minus' method of screening. If the clone of interest contains a sequence that is expressed only under certain conditions, probes may be made from mRNA populations from cells that are expressing the gene (the plus probe) and from cells that are not expressing the gene (the minus probe). By carrying out duplicate hybridisations, the clones can be identified by their different patterns of hybridisation with the plus and minus probes.

Genomic DNA probes are usually fragments of cloned sequences that are used either as heterologous probes or to identify other clones that contain additional parts of the gene in question. This is an important part of the techniques known as **chromosome walking** and **chromosome jumping**, and can enable the identification of overlapping sequences which, when pieced together, enable long stretches of DNA to be characterised.

The use of oligonucleotide probes is possible where some amino acid sequence data are available for the protein encoded by the target gene. Using the genetic code, the likely gene sequence can be derived and an oligonucleotide made. The degenerate nature of the genetic code means that it is not possible to predict the sequence with complete accuracy, but this is not usually a major problem, as mixed probes can be used that cover all the possible sequences. The great advantage of oligonucleotide probes is that only a short stretch of sequence is required for the probe to be useful, and thus genes for which clones are not already available can be identified by sequencing peptide fragments and constructing probes accordingly.

When a suitable probe has been obtained, it is usually labelled with ^{32}P as described in Section 3.3. This produces a radioactive fragment of high specific activity that can be used as an extremely sensitive screen for the gene of interest. Alternatively, non-radioactive labelling methods may be used if desired.

8.2.2 Screening clone banks

Colonies or plaques are not suitable for direct screening, so a replica is made on either nitrocellulose or nylon filters. This can be done either by growing cells directly on the filter on an agar plate (colonies), or by 'lifting' a replica from a plate (colonies or plaques). To do this the recombinants are grown and a filter is placed on the surface of the agar plate. Some of the cells/plaques

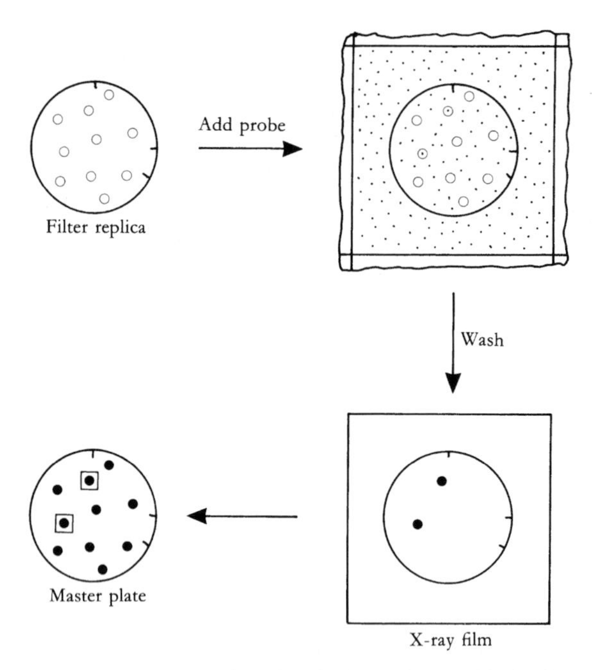

Fig. 8.3. Screening clone banks by nucleic acid hybridisation. A nitrocellulose or nylon filter replica of the master Petri dish containing colonies or plaques is made. Reference marks are made on the filter and the plate to assist with correct orientation. The filter is incubated with a labelled probe, which hybridises to the target sequences. Excess or non-specifically bound probe is washed off and the filter exposed to X-ray film to produce an autoradiograph. Positive colonies (boxed) are identified and can be picked from the master plate.

will stick to the filter, which therefore becomes a mirror image of the pattern of recombinants on the plate (Fig. 8.3). Reference marks are made so that the filters can be orientated correctly after hybridisation. The filters are then processed to denature the DNA in the samples, bind this to the filter and remove most of the cell debris.

The probe is denatured (usually by heating), placed in a sealed plastic bag with the filters, and incubated at a suitable temperature to allow hybrids to form. The **stringency** of hybridisation is important, and depends on conditions such as salt concentration and temperature. For homologous probes under standard conditions incubation is usually around 65–68 °C. Time of incubation may be up to 48 h in some cases, depending on the predicted kinetics of hybridisation. After hybridisation, the filters are washed (again the stringency of washing is important) and allowed to dry. They are then exposed to

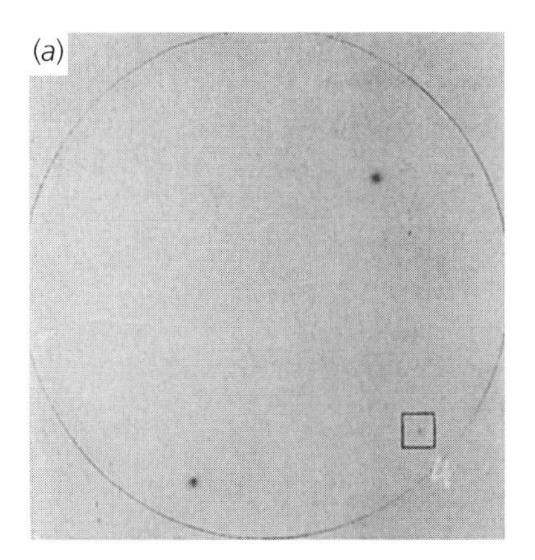

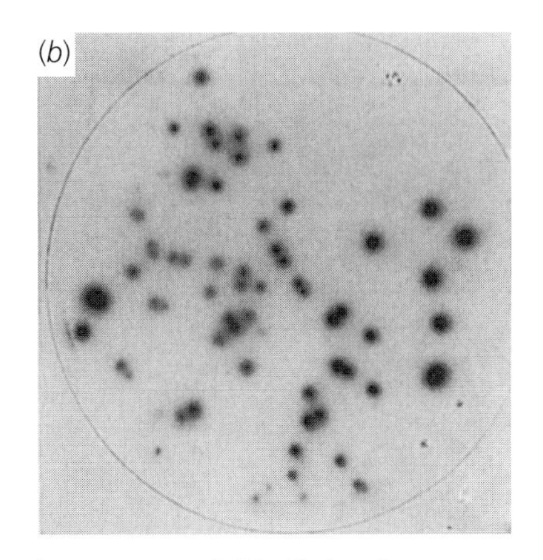

Fig. 8.4. Screening plaques at high and low densities. A radiolabelled probe was used to screen a genomic library in the λ vector EMBL3. (*a*) Initial screening was at a high density of plaques, which identified two positive plaques on this plate. The boxed area shows a false positive. (*b*) The plaques were picked from the positive areas and re-screened at a lower density to enable isolation of individual plaques. Many more positives are obtained due to the high proportion of 'target' plaques in the re-screened sample. (Photograph courtesy of Dr M. Stronach.)

X-ray film to produce an autoradiograph, which can be compared with the original plates to enable identification of the desired recombinant.

An important factor in screening genomic libraries by nucleic acid hybridisation is the number of plaques that can be screened on each filter. Often an initial high-density screen is performed, and the plaques picked from the plate. Because of the high plaque density, it is often not possible to avoid contamination by surrounding plaques. Thus the mixture is re-screened at a much lower plaque density, which enables isolation of a single recombinant (Fig. 8.4). This approach can be important if a large number of plaques has to be screened, as it cuts down the number of filters (and hence the amount of radioactive probe) required.

8.3 Immunological screening for expressed genes

An alternative to screening with nucleic acid probes is to identify the protein product of a cloned gene by immunological methods. The technique requires that the protein is expressed in recombinants, and is most often used for screening cDNA expression libraries that have been constructed in vectors such as λgt11. Instead of a nucleic acid probe, a specific antibody is used.

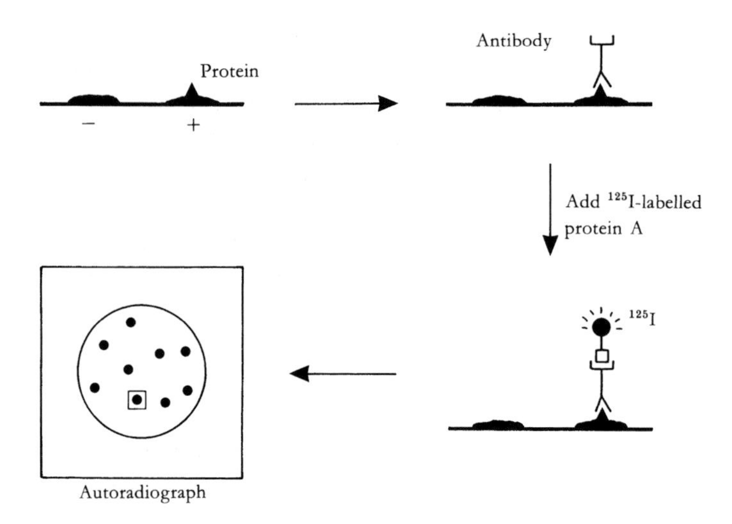

Fig. 8.5. Immunological screening for expressed genes. A filter is taken from a Petri dish containing the recombinants (usually cDNA/λ constructs). Protein and cell debris adhere to the filter. Plaques expressing the target protein (+) are indistinguishable from the others (−) at this stage. The filter is incubated with a primary antibody that is specific for the target protein. This is then complexed with radiolabelled protein A, and an autoradiograph prepared. As in nucleic acid screening, positive plaques can be identified and picked from the master plate. Chromogenic/enzymatic detection methods may also be used with this type of system.

Antibodies are produced by animals in response to challenge with an **antigen**, which is normally a purified protein. There are two main types of antibody preparation that can be used. The most common are **polyclonal** antibodies, which are usually raised in rabbits by injecting the antigen and removing a blood sample after the immune response has occurred. The immunoglobulin fraction of the serum is purified and used as the antibody preparation (antiserum). Polyclonal antisera contain antibodies which recognise all the antigenic determinants of the antigen. A more specific antibody can be obtained by preparing **monoclonal** antibodies, which recognise a single antigenic determinant. However, this can be a disadvantage in some cases. In addition, monoclonal antibody production is a complicated technique in its own right, and good quality polyclonal antisera are often sufficient for screening purposes.

There is a variety of methods available for immunological screening, but the technique is most often used in a similar way to 'plaque lift' screening with nucleic acid probes (Fig. 8.5). Recombinant λgt11 cDNA clones will express cloned sequences as β-galactosidase fusion proteins, assuming that the sequence is present in the correct orientation and reading frame. The proteins can be picked up onto nitrocellulose filters and probed with the antibody. Detection can be carried out by a variety of methods, most of which use a non-specific second binding molecule such as protein A from bacteria, or a second antibody, which attaches to the specifically bound primary antibody. Detection may be by radioactive label (^{125}I-labelled protein A or second antibody) or by non-radioactive methods which produce a coloured product.

8.4 Analysis of cloned genes

Once clones have been identified by techniques such as hybridisation or immunological screening, more detailed characterisation of the DNA can begin. There are many ways of tackling this, and the choice of approach will depend on what is already known about the gene in question, and on the ultimate aims of the experiment.

8.4.1 Characterisation based on mRNA translation *in vitro*

In some cases the identity of a particular clone may require confirmation. This is particularly true when the plus/minus method of screening has been used, as the results of such a process are usually somewhat ambiguous. If the desired sequence codes for a protein, and the protein has been characterised, it is possible to identify the protein product by two methods based on translation of mRNA *in vitro*. These methods are known as **hybrid-arrest translation** (**HART**) and **hybrid-release translation** (**HRT**). Although these techniques are now not widely used in gene analysis, they do illustrate how a particular problem can be approached by two different variations of a similar theme – a central part of good scientific method. A comparison of HART and HRT is shown in Fig. 8.6.

Both HART and HRT rely on hybridising cloned DNA fragments to mRNA prepared from the cell or tissue type from which the clones have been derived. In hybrid arrest, the cloned sequence blocks the mRNA and prevents its translation when placed in a system containing all the components of the translational machinery. In hybrid release, the cloned sequence is immobilised and used to select the clone-specific mRNA from the total mRNA preparation. This is then released from the hybrid and translated *in vitro*. If a radioactive amino acid (usually [^{35}S]methionine) is incorporated into the translation

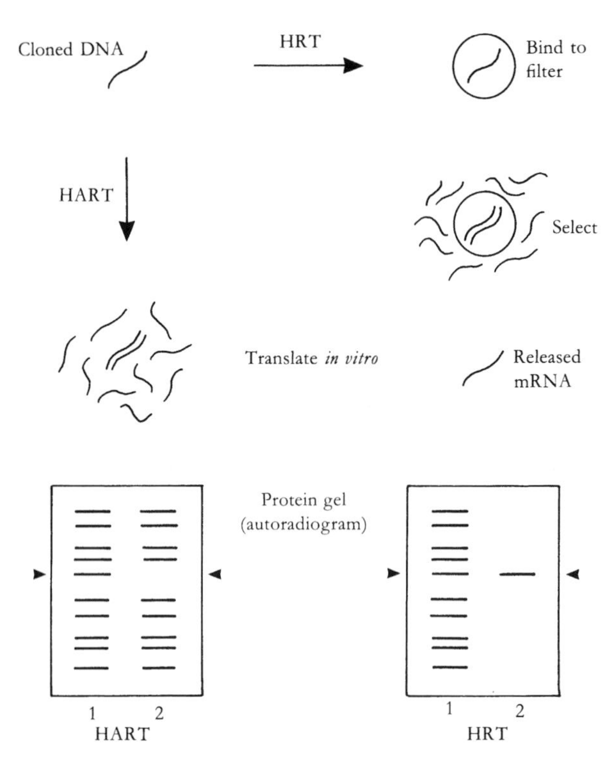

Fig. 8.6. Hybrid arrest and hybrid release translation to identify the protein product of a cloned fragment. In hybrid arrest (HART) the cloned fragment is mixed with a preparation of total mRNA. The hybrid formed effectively prevents translation of the mRNA to which the cloned DNA is complementary. After translation *in vitro*, the protein products of the translation are separated on a polyacrylamide gel. The patterns of the control (lane 1, HART gel) and test (lane 2, HART gel) translations differ by one band due to the absence of the protein encoded by the mRNA that has hybridised with the DNA. In hybrid release (HRT) the cloned DNA is bound to a filter and used to select its complementary mRNA from total mRNA. After washing to remove unbound mRNAs and releasing the specifically bound mRNA from the filter, translation *in vitro* generates a single band (lane 2, HRT gel) as opposed to the multiple bands of the control (lane 1, HRT gel). In both cases the identity of the protein (and hence the gene) can be determined by examination of the protein gels. The protein band of interest is arrowed.

mixture, the proteins synthesized from the mRNA(s) will be labelled and can be detected by autoradiography or fluorography after SDS–polyacrylamide gel electrophoresis. In hybrid arrest, one protein band should be absent, whilst in hybrid release there should be a single band. Thus hybrid release gives a cleaner result than hybrid arrest, and is the preferred method.

8.4.2 Restriction mapping

Obtaining a restriction map for cloned fragments is usually essential before additional manipulations can be carried out. This is particularly important where phage or cosmid vectors have been used to clone large pieces of DNA. If a restriction map is available, smaller fragments can be isolated and used for various procedures including sub-cloning into other vectors, the preparation of probes for chromosome walking, and DNA sequencing.

The basic principle of restriction mapping is outlined in Section 4.1.3. In practice, the cloned DNA is usually cut with a variety of restriction enzymes to determine the number of fragments produced by each enzyme. If an enzyme cuts the fragment at frequent intervals, it will be difficult to decipher the restriction map, so enzymes with multiple cutting sites are best avoided. Enzymes that cut the DNA into two to four pieces are usually chosen for initial experiments. By performing a series of single and multiple digests with a range of enzymes, the complete restriction map can be pieced together. This provides the essential information required for more detailed characterisation of the cloned fragment.

8.4.3 Blotting techniques

Although a clone may have been identified and its restriction map determined, this information in itself does not provide much of an insight into the fine structure of the cloned fragment and the gene that it contains. Ultimately, the aim may be to obtain the gene sequence (see Section 8.4.4), but it is usually not sensible to begin sequencing straight away. If, for example, a 20 kb fragment of genomic DNA has been cloned in a λ replacement vector, and the area of interest is only 2 kb in length, much effort would be wasted by sequencing the entire clone. In many experiments it is therefore essential to determine which parts of the original clone contain the regions of interest. This can be done by using a variety of methods based on blotting nucleic acid molecules onto membranes, and hybridising with specific probes. Such an

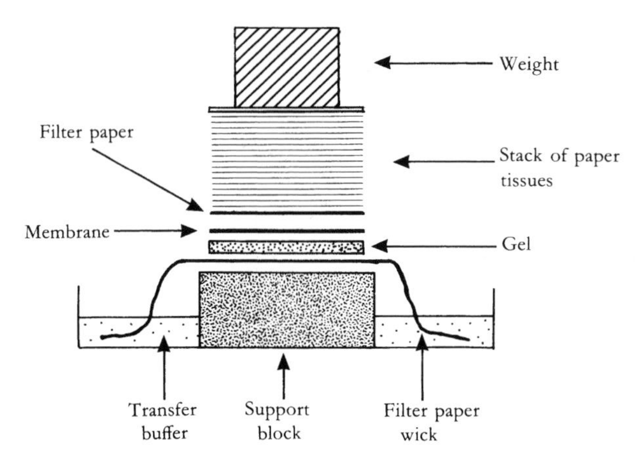

Fig. 8.7. Blotting apparatus. The gel is placed on a filter paper wick and a nitrocellulose or nylon filter placed on top. Further sheets of filter paper and paper tissues complete the set up. Transfer buffer is drawn through the gel by capillary action, and the nucleic acid fragments are transferred out of the gel and onto the membrane.

approach is in some ways an extension of clone identification by colony or plaque hybridisation, with the refinement that information about the structure of the clone is obtained.

The first blotting technique was developed by Ed Southern, and is eponymously known as **Southern blotting**. In this method fragments of DNA, generated by restriction digestion, are subjected to agarose gel electrophoresis. The separated fragments are then transferred to a nitrocellulose or nylon membrane by a 'blotting' technique. The original method used capillary blotting, as shown in Fig. 8.7. Although other methods such as vacuum blotting and electroblotting have been devised, the original method is still used extensively. Blots are often set up with whatever is at hand, and precarious-looking versions of the blotting apparatus are a common sight in many laboratories.

When the fragments have been transferred from the gel and bound to the filter, it becomes a replica of the gel. The filter can then be hybridised with a radioactive probe in a similar way to colony or plaque filters. As with all hybridisation, the key is the availability of a suitable probe. After hybridisation and washing, the filter is exposed to X-ray film and an autoradiograph prepared, which provides information on the structure of the clone. An example of the use of Southern blotting in clone characterisation is shown in Fig. 8.8.

Although Southern blotting is a very simple technique, it has many applications, and has been an invaluable method in gene analysis. The same technique can also be used with RNA, as opposed to DNA, and in this case is

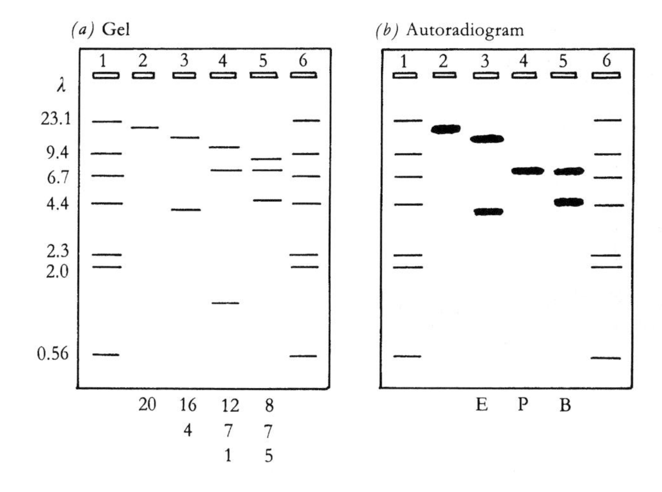

Fig. 8.8. Southern blotting. A hypothetical 20 kb fragment from a genomic clone is under investigation. A cDNA copy of the mRNA is available for use as a probe. (*a*) The gel pattern of fragments produced by digestion with various restriction enzymes; (*b*) the autoradiograph resulting from the hybridisation. Lanes 1 and 6 contain λ *Hin*dIII markers, sizes as indicated. These have been marked on the autoradiograph for reference. The intact fragment (lane 2) runs as a single band to which the probe hybridises. Lanes 3,4 and 5 were digested with *Eco*RI (E), *Pst*I (P) and *Bam*HI (B). Fragment sizes are indicated under each lane in (*a*). The results of the autoradiography show that the probe hybridizes to two bands in the *Eco*RI and *Bam*HI digests, therefore the clone must have internal sites for these enzymes. The *Pst*I digest shows hybridization to the 7 kb fragment only. This might therefore be a good candidate for sub-cloning, as the gene may be located entirely on this fragment.

known as **Northern blotting**. It is most useful in determining hybridisation patterns in mRNA samples, and can be used to determine which regions of a cloned DNA fragment will hybridise to a particular mRNA. However, it is more often used as a method of measuring transcript levels during expression of a particular gene.

There are two further variations on the blotting theme. If nucleic acid samples are not subjected to electrophoresis, but are spotted onto the filters, hybridisation can be carried out as for Northern and Southern blots. This technique is known as **dot-blotting**, and is particularly useful in obtaining quantitative data in the study of gene expression. The final technique is known as **Western blotting**, and this involves the transfer of electrophoretically separated protein molecules to membranes. The membrane is then probed with an antibody to detect the protein of interest, in a similar way to immunological screening of plaque lifts from expression libraries.

8.4.4 DNA sequencing

The development of rapid methods for sequencing DNA, as outlined in Section 3.6, has meant that this task has now become routine practice in most laboratories where cloning is carried out. Sequencing a gene provides much useful information about coding sequences, control regions and other features such as intervening sequences. Thus full characterisation of a gene will inevitably involve sequencing, and a suitable strategy must be devised to enable this to be achieved most efficiently. The complexity of a sequencing strategy depends on a number of factors, the main one being the length of the fragment that is to be sequenced. Most manual sequencing methods enable about 300–400 bases to be read from a sequencing gel. If the DNA is only a few hundred base-pairs long, it can probably be sequenced in a single step. However, it is more likely that the sequence will be several kilobase-pairs in length, and thus sequencing is more complex.

There are basically two ways of tackling large sequencing projects. Either a random or 'shotgun' approach is used, or an ordered strategy is devised in which the location of each fragment is known prior to sequencing. In the shotgun method, random fragments are produced and sequenced. Assembly of the complete sequence relies on there being sufficient overlap between the sequenced fragments to enable computer matching of sequences from the raw data.

An ordered sequencing strategy is usually more efficient than a random fragment approach. There are several possible ways of generating defined fragments for sequencing. Examples include: (i) isolation and sub-cloning of defined restriction fragments and (ii) generation of a series of sub-clones in which the target sequence has been progressively deleted by nucleases. If defined restriction fragments are used, the first requirement is for a detailed restriction map of the original clone. Using this, suitably sized fragments can be identified and sub-cloned into a sequencing vector such as M13 or pBluescript. Each sub-clone is then sequenced, usually by the dideoxy method (see Section 3.6.2). Both strands of the DNA should be sequenced independently, so that any anomalies can be spotted and re-sequenced if necessary. The complete sequence is then assembled by using a suitable computer software package. This is made easier if overlapping fragments have been isolated for sub-cloning, as the regions of overlap enable adjoining sequences to be identified easily.

By devising a suitable strategy and paying careful attention to detail, it is possible to derive accurate sequence data from most cloned fragments. The task of sequencing a long stretch of DNA is not trivial, but it is now such an integral part of gene manipulation technology that most gene-cloning projects involve sequence determination at some point. The technology has improved greatly to the point where entire genomes are being sequenced, which has ushered in a whole new era of molecular genetics. This aspect of sequencing will be discussed in the next chapter.

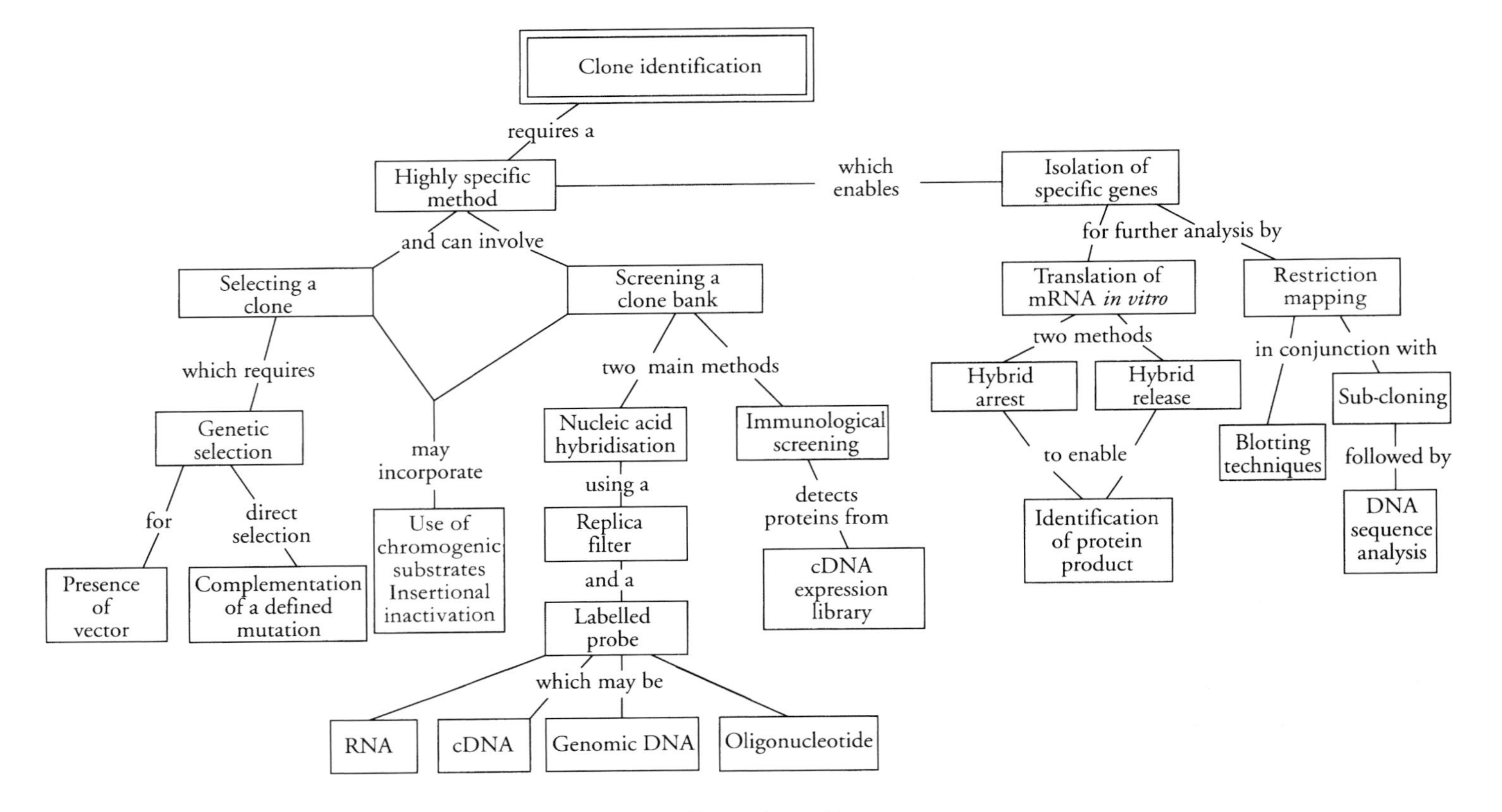

Concept map 8

Part III

Genetic engineering in action

9

Understanding genes and genomes

In Part II we have examined some of the basic techniques of gene manipulation. These techniques, and many more sophisticated variations of them, give the scientist the tools that enable genes to be isolated and characterised. In this final section of the book we will consider some of the applications of gene manipulation. Of necessity, this will be a highly selective treatment, the aim being to give some idea of the immense scope of the subject whilst trying to include some detail in certain key areas. We will also look more broadly at some of the ethical problems that gene manipulation poses, and at the topic of organismal cloning.

In many ways genetic engineering has undergone a shift in emphasis over the past few years, away from the technical problems that had to be solved before the technology became 'user friendly' enough for widespread use. Gene manipulation is now used as a tool to address many diverse biological problems that were previously intractable, and the applications of the subject appear at times to be limited only by the imagination of the scientists who use the technology in basic research, medicine, biotechnology and other related disciplines.

9.1 Analysis of gene structure and function

In terms of 'pure' science, the major impact of gene manipulation has been in the study of gene structure and expression. The organisation of genes within genomes is a fast-developing area that is essentially an extension of the early work on gene structure. Although the contribution of classical genetic analysis should not be underestimated, much of the fine detail regarding gene structure and expression remained a mystery until the techniques of gene cloning enabled the isolation of individual genes.

As discussed in Section 8.4, many of the techniques used to characterise cloned DNA sequences provide information about gene structure, with one of the aims of most experiments being the determination of the gene sequence. However, even when the sequence is available, there is still much work to be done to interpret the various structural features of the sequence in the context of their function *in vivo*. In this section I extend the discussion of gene analysis to include some of the methods used to investigate gene structure and function. We will also examine how modern gene manipulation technology has opened up the world of the genome in a way that was not thought possible a few years ago.

9.1.1 A closer look at sequences

Computer analysis of DNA sequences can provide much useful information about the structure and organisation of genes. Computers are ideally suited to this task of sequence analysis, which requires that fairly simple (but repetitive) operations are carried out quickly and accurately. Even a short sequence is tedious to analyse without the help of a computer, and there is always the possibility of error due to misreading of the sequence or to a loss in concentration. The whole area of DNA sequence analysis is often called **bioinformatics**, particularly in the context of understanding genome structure.

When a gene sequence has been determined, a number of things can be done with the information. Searches can be made for regions of interest, such as promoters, enhancers, etc., and for sequences that code for proteins. Coding regions can be translated to give the amino acid sequence of the protein. Restriction maps can be generated easily, and printed in a variety of formats. The sequence can be compared with others from different organisms and the degree of homology between them may be determined, which can assist in studying the phylogenetic relationships between groups of organisms.

Although computer analysis of a sequence is a very useful tool, it usually needs to be backed up with experimental evidence of structure or function. For example, if a previously unknown gene is being characterised, it will be necessary to carry out experiments to determine where the important regions of the gene are. Usually such experiments confirm the function inferred from the sequence analysis, although sometimes new information is generated. Thus it is important that the computational and experimental sides of sequence analysis are used in concert.

9.1.2 Finding important regions of genes

One of the key aspects in the control of gene expression concerns protein/DNA interactions. Thus it is important to find the regions of a sequence to which the various types of regulatory proteins will bind. A relatively simple way to do this is to prepare a restriction map of the cloned DNA and generate a set of restriction fragments. The protein under investigation (perhaps RNA polymerase, a repressor protein or some other regulatory molecule) is added and allowed to bind to its site. If the fragments are then subjected to electrophoresis, the DNA/protein hybrid will run more slowly than a control fragment without protein, and can be detected by its reduced mobility. This technique is known as **gel retardation**, and it provides information about the location of particular binding sites on DNA molecules.

Although gel retardation is a useful technique, its accuracy is limited by the precision of the restriction map and the sizes of fragments that are generated. A much more precise way of identifying regions of protein binding is the technique of **DNA footprinting** (sometimes called the **DNase protection** method). The technique is elegantly simple, and relies on the fact that a region of DNA that is complexed with a protein will not be susceptible to attack by DNase I (Fig. 9.1). The DNA fragment under investigation is radiolabelled and mixed with a suspected regulatory protein. DNase I is then added so that limited digestion occurs; on average, one DNase cut per molecule is achieved. Thus a set of nested fragments will be generated, and these can be run on a sequencing gel. The region that is protected from DNase digestion gives a 'footprint' of the binding site within the molecule.

It is often necessary to locate the start site of transcription for a particular gene, and this may not be apparent from the gene sequence data. Two methods can be used to locate the T_C start site, these being **primer extension** and **S_1 mapping**. In primer extension a cDNA is synthesised from a primer that hybridises near the 5′ end of the mRNA. By sizing the fragment that is produced, the 5′ terminus of the mRNA can be identified. If a parallel sequencing reaction is run using the genomic clone and the same primer, the T_C start site can be located on the gene sequence. In S_1 mapping the genomic

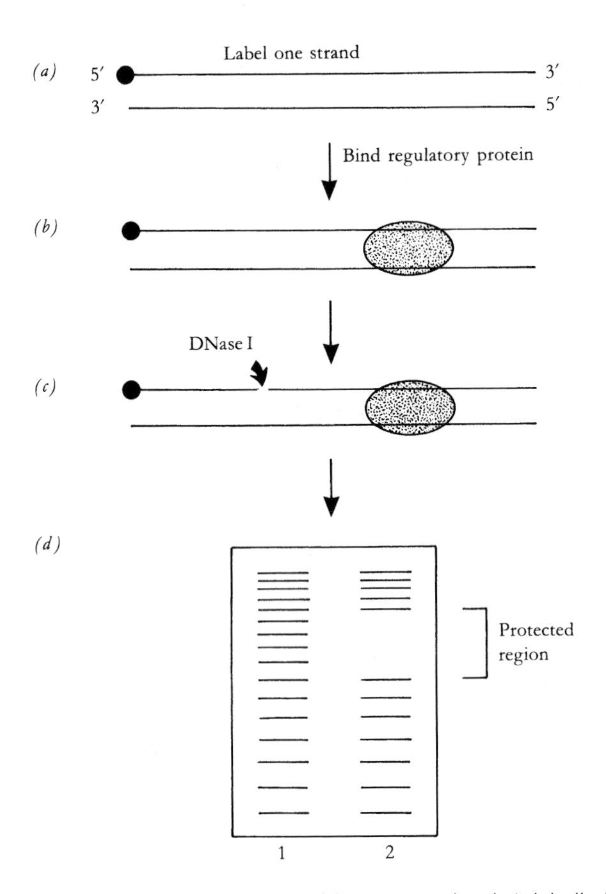

Fig. 9.1. DNA footprinting. (*a*) A DNA molecule is labelled at one end with ^{32}P. (*b*) The suspected regulatory protein is added and allowed to bind to its site. A control reaction without protein is also set up. (*c*) DNase I is used to cleave the DNA strand. Conditions are chosen so that on average only one nick will be introduced per molecule. The region protected by the bound protein will not be digested. Given the large number of molecules involved, a set of nested fragments will be produced. (*d*) The reactions are then run on a sequencing gel. When compared to the control reaction (lane 1), the test reaction (lane 2) indicates the position of the protein on the DNA by its 'footprint'.

fragment that includes the T_C start site is labelled and used as a probe. The fragment is hybridised to the mRNA and the hybrid then digested with single-strand-specific S_1-nuclease. The length of the protected fragment will indicate the location of the T_C start site relative to the end of the genomic restriction fragment.

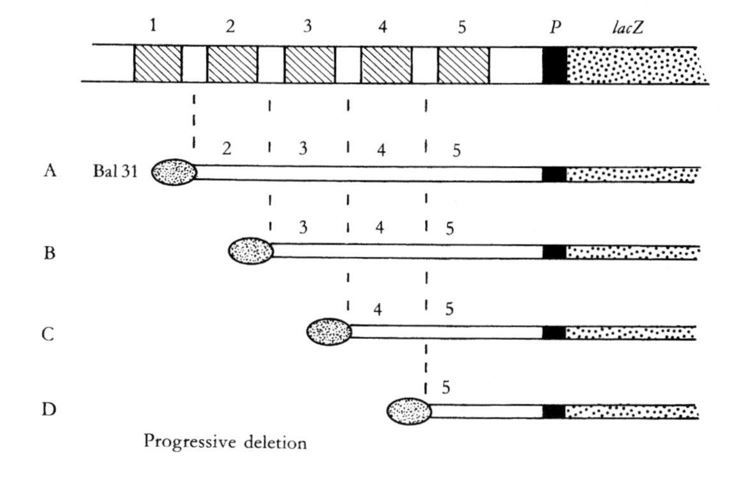

Fig. 9.2. Deletion analysis in the study of gene expression. In this hypothetical example a gene has five suspected upstream controlling regions (1 to 5, hatched). The gene promoter is labelled P. Often the *lacZ* gene is used as a reporter gene for detection of gene expression using the X-gal system. Deletions are created using an enzyme such as Bal 31 nuclease. In this example four deletion constructs have been made (labelled A to D). In A, region 1 has been deleted, with progressively more upstream sequence removed in each construct so that in D regions 1,2,3 and 4 have been deleted and it retains only region 5. The effects of these deletions can be monitored by the detection of β-galactosidase activity, and thus the positions of upstream controlling elements can be determined. As an alternative to using Bal 31, restriction fragments can be removed from the controlling region.

9.1.3 Investigating gene expression

Recombinant DNA technology can be used to study gene expression in two main ways. Firstly, genes that have been isolated and characterised can be modified and the effects of the modification studied. Secondly, probes that have been obtained from cloned sequences can be used to determine the level of mRNA for a particular protein under various conditions. These two approaches, and extensions of them, have provided much useful information about how gene expression is regulated in a wide variety of cell types.

One method of modifying genes to determine which regions are important in controlling gene expression is to delete sequences lying upstream from the T_C start site. If this is done progressively using a nuclease such as exonuclease III or Bal 31 (see Section 4.2.1), a series of deletions is generated (Fig. 9.2). The effects of the various deletions can be studied by monitoring the level of

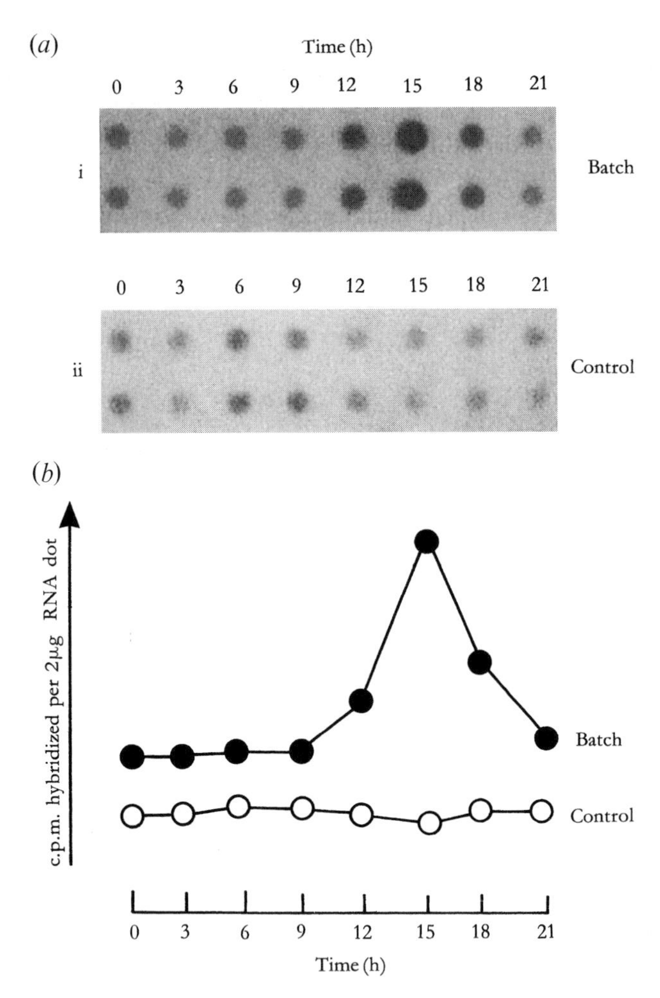

Fig. 9.3. Dot-blot analysis of mRNA levels. Samples of total RNA from synchronous cell cultures of *Chlamydomonas reinhardtii* grown under batch culture and turbidostat (control) culture conditions were spotted onto a membrane filter. The filter was probed with a radiolabelled cDNA specific for an mRNA that is expressed under conditions of flagellar regeneration. (*a*) An autoradiograph was prepared after hybridisation. Batch conditions (i) show a periodic increase in transcript levels with a peak at 15 h. Control samples (ii) show constant levels. Data shown in (*b*) were obtained by counting the amount of radioactivity in each dot. This information can be used to determine the effect of culture conditions on expression of the flagellar protein. Photograph courtesy of Dr J. Schloss. From Nicholl *et al*. (1988), *Journal of Cell Science* **89**, 397–403. Copyright (1988) The Company of Biologists Limited. Reproduced with permission.

expression of the gene itself, or of a 'reporter' gene such as the *lacZ* gene. In this way regions that increase or decrease transcription can be located, although the complete picture may be difficult to decipher if multiple control sequences are involved in the regulation of transcription.

Measurement of mRNA levels is an important aspect of studying gene expression, and is often done using cDNA probes that have been cloned and characterised. The mRNA samples for probing may be from different tissue types or from cells under different physiological conditions, or may represent a time-course if induction of a particular protein is being examined. If the samples have been subjected to electrophoresis a Northern blot can be prepared, which gives information about the size of transcripts as well as their relative abundance. Alternatively, a dot-blot can be prepared and used to provide quantitative information about transcript levels by determining the amount of radioactivity in each 'dot': this reflects the amount of specific mRNA in each sample (Fig. 9.3).

Northern and dot-blotting techniques can provide a lot of useful data about transcript levels in cells under various conditions. When considered along with information about protein levels or activities, derived from Western blots or enzyme assays, a complete picture of gene expression can be built up.

9.2 From genes to genomes

Although much emphasis is still placed on the analysis of individual genes, advances in gene manipulation technology have opened up the study of genomes to the point where this is emerging as a discipline in its own right. Often called simply **genomics**, the emphasis here is on a holistic approach to how genomes function. We are therefore now much more likely to assess the function of a gene within the context of its role in the genome, as opposed to considering gene structure and expression in isolation.

Genes themselves are of course only the starting point for the study of how genetics enables cells to perform all the various functions that are required. The complementary area of how proteins function is now being re-examined in the light of more information about genomes. The term **proteome** is used to describe the set of proteins encoded by the genetic information in a cell, and great advances are being made in this area of research. The concept of the cell as a **molecular machine** is one that goes some way to describing the emphasis of modern molecular biology, in which genes are just part of the overall picture. The biochemical interactions between the various proteins

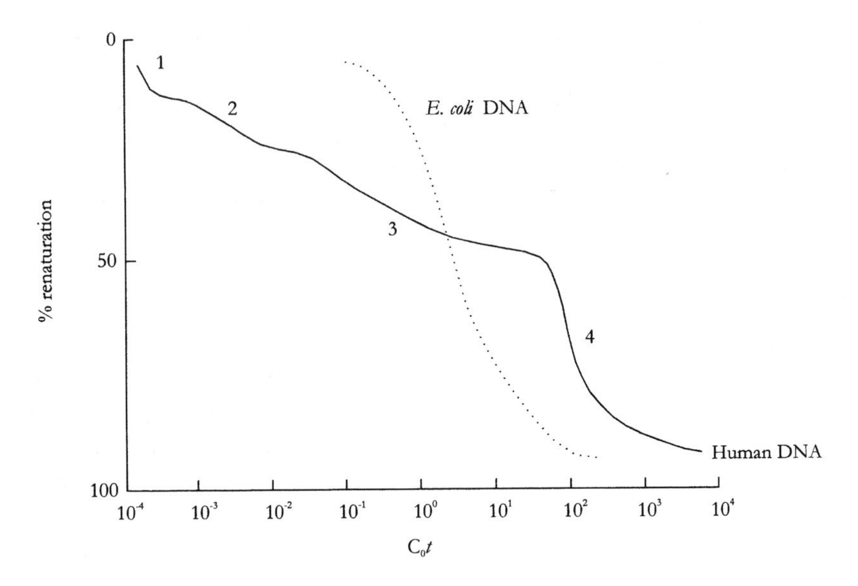

Fig. 9.4. Abundance classes found in different genomic DNA samples. Following denaturation, samples of DNA are allowed to cool and the % renaturation measured. This is plotted against log C_0t (this is functionally a measure of time, with the added context of DNA concentration). The dotted line shows *E. coli* DNA, with a simple one-step pattern. Human DNA (solid line) shows a complex pattern with four classes recognised: (1) foldback DNA, (2) highly repetitive sequences, (3) moderately repetitive sequences, and (4) unique or single-copy sequences.

of the cell are just as important as the information in the genes themselves, and the ultimate aim of cell and molecular biology is to understand cells fully at the molecular level. In many ways this is the biologist's holy grail, similar in scope to the search in physics for the unifying theory that will link the various branches of the discipline together.

9.2.1 Analysing genomes

The analysis of a genome is obviously much more complex than the analysis of a gene sequence, and involves a variety of techniques. Structural features such as repetitive sequence elements and intervening sequences can make the task much more difficult than might be supposed when simply considering the number of base-pairs, even though this is a major factor in determining the strategy for studying a particular genome. Some comparative genome sizes were presented in Table 2.3; these data show that eukaryotic genomes may be several orders of magnitude larger than bacterial genomes.

Some of the earliest indications of genome complexity were obtained by using the technique of **renaturation kinetics**. A sample of DNA is heated to denature the DNA. The strands are then allowed to re-associate as the mixture cools, and the UV-absorbance (A_{260}) is monitored. Single-stranded DNA has a higher absorbance than double-stranded DNA (this is called the **hyperchromic** effect), so the degree of renaturation can be assessed easily. Using this type of analysis, eukaryotic DNA was shown to be composed of several different **abundance classes**, as shown in Fig. 9.4. In the case of human DNA, about 40% of the total is either highly or moderately **repetitive sequence DNA**, which can often cause problems in the cloning and analysis of genes. Of the remaining 60%, which represents unique sequence and low copy number sequence elements, only around 3% is the actual coding sequence. This immediately poses a problem in the analysis of the human genome, in that 97% of the DNA could perhaps be 'avoided' if the genes themselves could be identified for further study.

Although some useful knowledge about genomes can be obtained by using kinetic analysis and identifying individual genes, the amount of detailed information provided by these methods is relatively small. The ultimate aim of genome analysis is determination of the DNA sequence, a task which represents a major undertaking even for a small genome. However, in recent years great advances have been made in the technology of DNA sequencing, and genome sequencing is now well established for many organisms. We will discuss the human genome in more detail in Section 9.3, after we have examined some of the methods for mapping and sequencing genomes.

9.2.2 Mapping genomes

Large-scale DNA sequencing can be done using a 'shotgun' method in which random fragments are sequenced. The sequences are then pieced together by matching overlaps. Despite its simplicity, this approach is rather limited, and most sequencing projects involving large genomes are carried out using an ordered strategy involving **genetic mapping** and **physical mapping** before sequence analysis is attempted. This approach requires access to large clone banks (YAC or BAC vectors are often used) and a well-developed system for recording the information that is generated. An analogy may help to put the whole process in context.

Genome mapping is a bit like using a road map. Say you wish to travel by car from one city to another – perhaps Glasgow to London, or New York to Denver. You could (theoretically) get hold of all the street maps for towns that exist along a line between your start and finish points, and follow these. However, there would be gaps, and you might go in the wrong direction from time to time. A much more sensible strategy would be to look at a map that showed the whole of the journey, and pick out a route with major landmarks – perhaps other cities or major towns along the way. At this initial planning stage of your journey you would not be too concerned about accurate distances – but you would like to know which roads to take. You would then split the journey into stages where local detail becomes very important – you need to know how far it is to the next petrol station, for example. For a stopover, the address of a hotel would guide you to the precise location. In genome analysis, the techniques of genetic and physical mapping provide the equivalent of large- and small-scale maps to enable progress to be made. Determination of the sequence itself is the level of detail that enables precise location of genome 'landmarks' such as genes and control regions.

Genetic mapping has provided a lot of information about the relative positions of genes on the chromosomes of organisms that can be used to set up experimental genetic crosses. The technique is based on the analysis of **recombination frequency** during meiosis, and is often called **linkage mapping**. This approach relies on having **genetic markers** that are detectable. The marker alleles must be heterozygous so that meiotic recombination can be detected. If two genes are on different chromosomes, they are unlinked and will assort independently during meiosis. However, genes on the same chromosome are physically linked together, and a crossover between them during prophase I of meiosis can generate non-parental genotypes. The chance of this happening depends on how far apart they are – if very close together, it is unlikely that a crossover will occur between them. If far apart, they may behave as though they are essentially unlinked. By working out the recombination frequency it is therefore possible to produce a map of the relative locations of the marker genes.

Physical mapping of genomes builds on genetic mapping and adds a further level of detail. As with genetic maps, construction of a physical map requires markers that can be mapped to a specific location on the DNA sequence. The aim of a physical map is to cover the genome with these identifiable **physical markers** that are spaced appropriately. If the markers are too far apart, the map will not provide sufficient additional information to be useful. If markers are not spread across the genome, there may be sections that have too few markers, whilst others have more than might be required for that particular stage of the investigation.

Table 9.1. *Some methods for physical mapping of genomes*

Technique name	Application
Clone mapping	Define the order of cloned DNA fragments by matching up overlapping areas in different clones. This generates a set of contiguous clones known as contigs. Clone mapping may be used with large or small cloned fragments as appropriate.
Radiation hybrid mapping	Fragment the genome into large pieces and locate markers on the same piece of DNA. The technique requires rodent cell lines to construct hybrid genomes as part of the process.
Fluorescent *in situ* hybridisation (FISH)	Locate DNA fragments by hybridisation, plotting the chromosomal position of the sequence by analysing the fluorescent markers used.
Long-range restriction mapping	Method is similar to any other restriction mapping procedure, but enzymes that cut infrequently in the DNA are used to enable long-range maps to be constructed.
Sequence-tagged site (STS) mapping	The most powerful technique, which can complement the other techniques of genetic and physical mapping. Can be applied to any part of the DNA sequence, as long as some sequence information is available. An STS is simply a unique identifiable region in the genome.
Expressed sequence tag (EST) mapping	A variant of STS mapping in which expressed sequences (i.e. gene sequences) are mapped. More limited than STS mapping, but useful in that genes are located by this method.

Physical mapping of genomes is not a trivial task, and until the early 1980s it was thought that a physical map of the human genome was unlikely to be achieved. However, this proved to be incorrect, and techniques for physical mapping of genomes were developed relatively quickly. Physical maps of the genome can be constructed in a number of ways, all of which have the aim of generating a map in which the distances between markers are known with reasonable accuracy. The various methods that can be used for physical mapping are shown in Table 9.1. **Sequence-tagged site** (**STS**) mapping has become the most useful method, as it can be applied to any area of sequence that is unique in the genome and for which DNA sequence information is available.

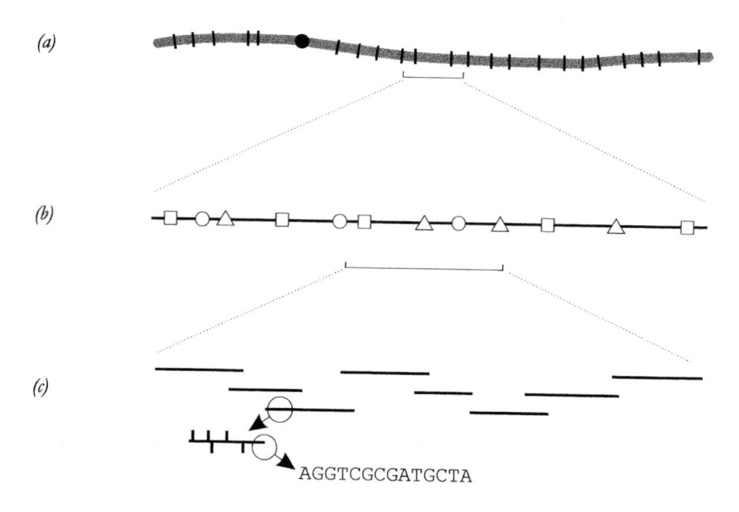

Fig. 9.5. Genome mapping. In (*a*) the genetic map is shown, with genetic markers assigned to positions on the chromosome. In (*b*) a section of the chromosome is shown, with the physical map of this region. Different types of physical marker are represented as different shapes. Various methods may be used to assign physical markers to their chromosomal locations (STS, FISH, etc., see Table 9.1). In (*c*) the clone map of a section of the physical map is shown, with large overlapping DNA fragments. From this a more detailed restriction map and the DNA sequence itself can be determined. From Nicholl (2000), *Cell and Molecular Biology*, Advanced Higher Monograph Series, Learning and Teaching Scotland. (Reproduced with permission.)

One advantage of STS mapping is that it can tie together physical map information with that generated by other methods. The technique essentially uses the sequence itself as the marker, identifying it either by hybridisation techniques or, more easily, by amplifying the sequence using PCR. With regard to the human genome, STS mapping (with other methods) enabled the construction of a useful genetic and physical map of the human genome by the late 1990s.

The final level of detail in genome mapping is usually provided by some sort of restriction map (see Section 4.1.3 for an illustration of restriction mapping). Long-range restriction mapping using enzymes that cut infrequently is a useful physical mapping technique (Table 9.1), but more detailed restriction maps are required when cloned fragments are being analysed prior to sequencing. The overall link between genetic, physical and clone maps is shown in Fig. 9.5.

9.3 Genome sequencing

The final stage of any large-scale sequencing project (such as a genome project) is to determine and assemble the actual DNA sequence itself. There are several critical requirements for this part. The **DNA sequencing technology** has to be accurate and fast enough to do the job in the proposed timescale, and there must be a library of **cloned fragments** available for sequencing. These two requirements will differ in scale for different projects.

9.3.1 Sequencing technology

When considering sequence determination itself, it was apparent from the start of large-scale projects that the traditional laboratory methods could not deliver the rate of progress that was required to complete the task in a sensible timescale. Thus automated sequencing methods were developed, in which the standard chain-termination method of sequencing was adapted by using **fluorescent labelling** instead of radioactive methods. In one method, the ddNTPs can be tagged with different labels, and one reaction carried out where all four ddNTPs are used together. The products are separated by gel electrophoresis, and the fluorescent labels detected as they come off the bottom of the gel. This gives a direct readout of the sequence. The process is much faster than conventional sequencing, and can be run continuously to provide large amounts of data in a short time. It is summarised in Fig. 9.6.

For large sequencing projects such as the human genome, an **integrated strategy** is required, with many centres being involved, each with specific jobs to do. Finally, appropriate **information technology** is required to handle the bioinformatics side of sequencing projects. These days, much of the information is published on the World Wide Web, where it is immediately available to researchers around the world.

9.3.2 Genome projects

The technology for large-scale DNA sequencing has enabled scientists to undertake genome sequencing in a realistic timescale. The roots of genome sequencing go back to 1983, when the sequence of bacteriophage λ was published. This was the first 'large' sequence to be completed – 48 502 base-pairs. Since then there has been a lot of activity in the mapping and sequencing of genomes. In the next section we will look at the human genome project, but

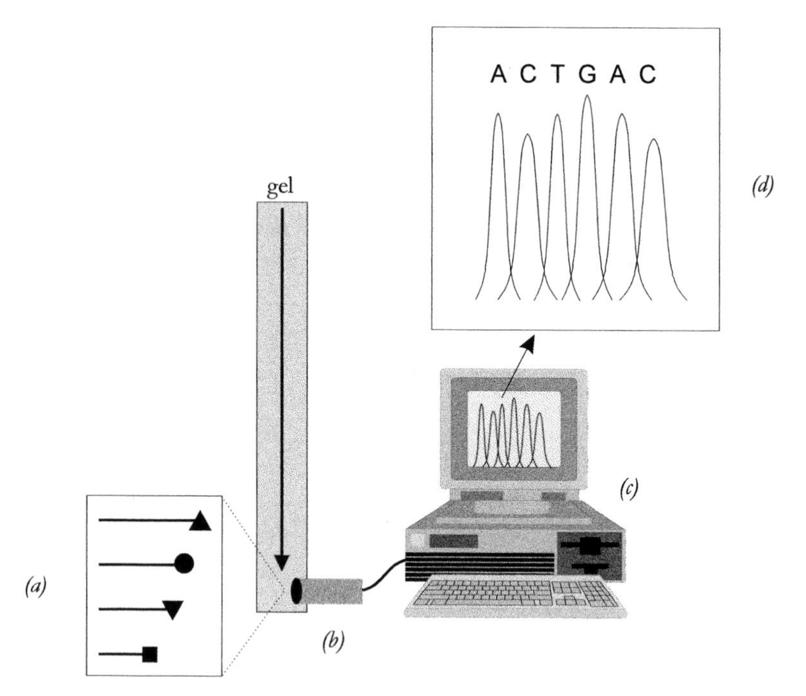

Fig. 9.6. Automated DNA sequencing using fluorescent marker dyes. Each ddNTP is tagged with a different dye, which generates a set of differently labelled nested fragments as shown in (*a*). On separation in a single lane of a sequencing gel, the DNA fragments pass through a detector (*b*) and the fluorescent labels are monitored. A computer captures the data and displays the sequence as a series of peaks (*c*), from which the sequence is read as shown in (*d*). From Nicholl (2000), *Cell and Molecular Biology*, Advanced Higher Monograph Series, Learning and Teaching Scotland. Reproduced with permission.

there are many other projects that have either been partially or fully completed. In fact, the number of different projects, and the enormous amount of effort that is put into these, illustrates how the emphasis in sequence analysis is changing from genes to genomes.

How are different organisms selected for sequencing? In most cases there is already a well-established history of research, and many 'model organisms' have a worldwide community of scientists working on many diverse aspects of their molecular biology, biochemistry, physiology and ecology. The range covers many different groups – bacteria, slime moulds, yeasts, nematodes, fruit flies, plants and mammals. Some notable examples include the bacterium *Escherichia coli*, the yeast *Saccharomyces cerevisiae*, the nematode worm

Caenorhabditis elegans, the fruit fly *Drosophila melanogaster*, the plant (some call it a weed!) *Arabidopsis thaliana* and the mouse *Mus musculus*. Naturally, molecular biologists working with these organisms want to sequence their genomes, to provide detailed information about genes, their organisation in the genome, and how they are expressed. Some non-human genome sequencing projects are listed in Table 9.2.

Different levels of 'completeness' can be recognised in any genome project, and it is perhaps a little unrealistic to say that any project is 'finished' – in fact, the complete sequence is in many ways just the beginning, and understanding how the genome functions is a task that will occupy scientists for many years to come. Major milestones in genome projects include the establishment of the genetic and physical maps, the production of unfinished or first draft sequence, and confirmation of the final version that is accepted as being the most accurate sequence possible. This usually involves sequencing each part of the genome several times, and cross-referencing the sequences to ensure that any gaps are filled in and sequence anomalies are cleared up.

In addition to deciphering the genetics of any particular organism, genome sequencing also opens up the field of **comparative genome analysis**, which can help to understand how genomes evolve and how many genes are similar in all organisms. Thus humans have genomes that are only about 0.1% different between individuals, 2% different from the chimpanzee, and which contain many of the same genes as the bacterial genome. Given such a wide range of interests, the future for bioinformatics, genomics and proteomics looks to be secure, and there will undoubtedly be many surprises in store as we seek to understand how cells and organisms function by examining their genetic information in detail.

9.4 The human genome project

Worms, weeds and fruitflies are all very well, but public attention is, of course, usually focused on any developments that involve *Homo sapiens*. Over the past decade or so, the major problems of human genome mapping and sequencing have been addressed, and the Human Genome Project (HGP) is now well established. The date of Monday 26th June 2000 will be noted as perhaps one of the most important dates in our history, as this was the date on which the 'first draft' of the human genome sequence was announced. Although this is of course a largely symbolic date (for reasons already discussed), it does rank alongside the discovery of the double helix in 1953 as one of the most exciting events in recent history. These two dates, with DNA as the focus, serve to

Table 9.2. *Selected genome sequencing projects*

Organism (type)	Sequence completed	Information	Relevant website(s)
Escherichia coli (bacterium)	1997	Genome size 4.64 Mb, which encodes 4405 genes	http://www.genome.wisc.edu/
Bacillus subtilis (bacterium)	1997	First gram-positive bacterial genome to be sequenced	http://genolist.pasteur.fr/SubtiList/
Saccharomyces cerevisiae (yeast)	1996	First eukaryotic genome sequenced Genome size 12 Mb, estimated to contain around 6000 genes	http://www.mips.biochem.mpg.de/
Caenorhabditis elegans (nematode worm)	1998	First multicellular organism's genome to be sequenced. Genome size 100 Mb	http://www.wormbase.org/
Drosophila melanogaster (fruit fly)	2000	Shotgun approach used for this complex genome. Some 14000 genes in 165 Mb	http://flybase.bio.indiana.edu/
Arabidopsis thaliana (plant)	2000	Genome size 117 Mb, the first plant genome sequence to be determined	http://www.arabidopsis.org/ http://www.mips.biochem.mpg.de/
Mus musculus (mouse)	Ongoing	0.9% finished and 9% draft sequence at start of 2001	http://www.informatics.jax.org/

Note: Date refers to publication of complete genome sequence – this is usually a first draft sequence that is then updated as more finished sequence becomes available. Website URLs are correct as of January 2001. There are usually multiple websites for each genome project; these can usually be accessed from links pages in the sites listed. Alternatively, a search using [organism's name] and [genome sequence] (or some such combination of terms) will often get to relevant sites. A useful gateway site is the *Nature* Genome Gateway at http://www.nature.com/genomics.

illustrate how technology has changed science. Watson and Crick were part of a relatively small community of people who knew about DNA in the 1950s, and they constructed models using clamps and stands, and bases made in their workshop at Cambridge. In contrast, sequencing the genome has required a major international collaborative effort, with around 20 major sequencing centres worldwide, and many thousands of molecular biologists, technicians and computer scientists involved either directly or indirectly.

The idea that the human genome could be sequenced gained credibility in the mid 1980s, and by the end of the decade the project had acquired sufficient momentum to ensure that it would be supported. The initial impetus had come from the USA, and the Formation of the Human Genome Organisation (HUGO) in 1988 marked the birth of the project on an international scale, the role of HUGO being to co-ordinate the efforts of the many countries involved. The Human Genome Project (HGP) was officially launched in October 1990, and presented a task of almost unimaginable complexity and scale. Molecular biology has traditionally involved small groups of workers in individual laboratories, and most of the key discoveries have been made in this way. Sequencing the 3×10^9 base-pairs of the human genome is in another league altogether. The analogies shown in Table 9.3 may help put the scale of the project into some sort of context.

9.4.1 Whose genome, and how many genes does it contain?

A question that is often asked is 'whose genome is being sequenced, and how will this relate to the other 6 billion or so variations that exist?'. As might be expected, there is no simple answer to this. The positive aspect of this question is that it can largely be ignored – it does not really matter *whose* genome is sequenced, as the phenotypic differences between individuals are generated from very little overall variation in the sequence itself (around 0.1% – this still represents around 3 million base-pair differences between individuals!). Such differences or **polymorphisms** will in fact be one area that is of great interest when examining how the genome functions, and thus particular loci may be sequenced many times for different reasons. The reality is that many individual genomes are being used as the source material for mapping and sequencing studies.

The other question that has been the subject of much debate concerns the number of genes in the genome. The figure of around 100000 is often quoted, and is a reasonable number to use as a rough guide to the complexity of the genome. Estimates made in the early 1990s suggested around 50000

Table 9.3. *Some interesting facts about our genome*

The information would fill two hundred 500-page telephone directories.
Between humans, our DNA differs by only 0.2%, or 1 in 500 bases (letters). (This takes into account that human cells have two copies of the genome.)
If we recited the genome at one letter per second for 24 hours a day it would take a century to recite the book of life.
If two different people started reciting their individual books at a rate of one letter per second, it would take nearly eight and a half minutes (500 seconds) before they reached a difference.
A typist typing at 60 words per minute (around 360 letters) for 8 hours a day would take around 50 years to type the book of life.
Our DNA is 98% identical to that of chimpanzees.
The vast majority of DNA in the human genome – 97% – has no known function.
The first chromosome to be completely decoded was chromosome 22 at the Sanger Centre in Cambridgeshire, in December 1999.
There is 6 feet of DNA in each of our cells packed into a structure only 0.0004 inches across (it would easily fit on the head of a pin).
There are 3 billion (3 000 000 000) letters in the DNA code in every cell in your body.
If all the DNA in the human body was put end to end it would reach to the sun and back over 600 times (100 trillion × 6 feet divided by 93 million miles = 1200).

Note: Information taken from the Sanger Centre website [http://www.sanger.ac.uk]. Reproduced with permission. The Sanger Centre is supported by the Wellcome Trust.

genes, but this was revised upwards as more gene sequences were determined. Some people put the number as high as 150 000, although this is thought by most to be too many. By using a range of methods, the best estimate seemed to be around 70–80 000 genes, although at the time of writing (February 2001) an estimate of 30 000 was made that looks to be reasonably accurate. A website at URL [**http://www.ensembl.org/Genesweep/**] was taking bets on the actual number!

9.4.2 Genetic and physical maps of the human genome

In many ways the critical phase of the genome project was not the actual sequencing, but the genetic and physical mapping required to enable the sequence to be compiled against a reference map. Genetic mapping has been hampered by the fact that experimental crosses cannot be set up in humans, and therefore mapping must be a retrospective activity. In many cases tracing the inheritance pattern of genetic markers associated with disease can provide much useful information. The **Centre d'Étude Polymorphism Humain** (**CEPH**) in Paris maintains a set of reference cell lines extending over three generations (families of four grandparents, two parents and at least six children). These have been extremely useful in mapping studies.

Many thousands of diseases of genetic origin have been identified in which the defect is traceable as a **monogenic** (single gene) disorder. Currently, over 12000 entries (of various types) are listed in **Online Mendelian Inheritance in Man**, which is the central datastore site for this information. This can be found at URL [**http://www.ncbi.nlm.nih.gov/Omim/**]. Many of these diseases have already been studied extensively using the retrospective technique of **pedigree analysis**. However, to generate useful genetic map data, it is often not necessary to be able to trace the actual gene responsible for the phenotypic effect. If a polymorphic marker can be identified that almost always segregates with the target gene, this can be just as useful. These are called **neutral molecular polymorphisms**. In the early 1980s a marker of this group called a **restriction fragment length polymorphism** (**RFLP**) began to be used to map genes. RFLPs are differences in the lengths of specific restriction fragments generated when DNA is digested with a particular enzyme (Fig. 9.7). They are produced when there is a variation in DNA that alters either the recognition sequence or the location of a restriction enzyme recognition site. Thus a point mutation might abolish a particular restriction site (or create a new one), whereas an insertion or deletion would alter the relative positions of restriction sites. If the RFLP lies within (or close to) the locus of a gene that causes a particular disease, it is often possible to trace the defective gene by looking for the RFLP, using the Southern blotting technique in conjunction with a probe that hybridizes to the region of interest. This approach is extremely powerful, and enabled many genes to be mapped to their chromosomal locations before high-resolution genetic and physical maps became available. Examples include the genes for Huntington disease (chromosome 4), cystic fibrosis (chromosome 7), sickle-cell anaemia (chromosome 11), retinoblastoma (chromosome 13) and Alzheimer disease (chromosome 21).

Using RFLPs as markers, a genetic map of the human genome was available by 1987. However, this approach was limited in terms of the degree of polymorphism (a restriction site can only be present or absent) and in the level of resolution. The use of **minisatellites** and **microsatellites** enabled more detailed genetic maps to be constructed. Minisatellites are made up of tandem

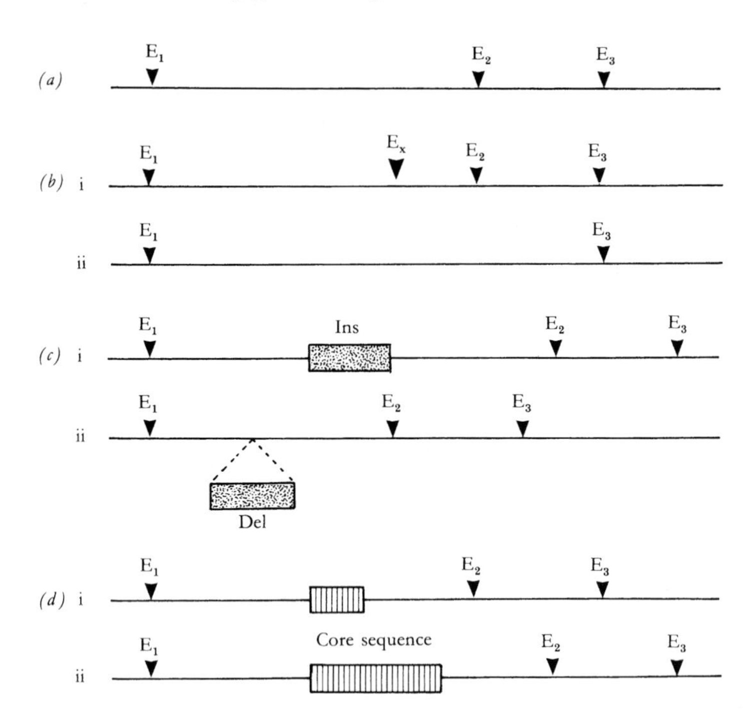

Fig. 9.7. Possible ways of generating restriction fragment length polymorphisms (RFLPs). (*a*) Consider a DNA fragment with three *Eco*RI sites (E_1, E_2 and E_3). On digestion with *Eco*RI, one of the fragments produced is E_1 E_2. RFLPs can be generated if the relative positions of these two sites are altered in any way. (*b*) The effect of point mutations. If a point mutation creates a new *Eco*RI site, ((*b*)i, marked E_X), fragment E_1 E_2 is replaced with two shorter fragments, E_1 E_X and E_X E_2. If a point mutation removes an *Eco*RI site ((*b*)ii, site E_2 removed), the fragment becomes E_1 E_3, which is longer than the original fragment. (*c*) The effect of insertions or deletions. If additional DNA is inserted between E_1 and E_2 (Ins in (*c*)i), fragment E_1 E_2 becomes larger. Insertions might also carry additional *Eco*RI sites, which would affect fragment lengths. If DNA is deleted (Del in (*c*)ii), the fragment is shortened. (*d*) The effect of variable numbers of repetitive core sequence motifs. This variation can be considered as a type of RFLP, which can be used as the basis of genetic profiling. (*d*)i has 10 copies of the repeated core sequence element, thus fragment E_1 E_2 is smaller than that shown in (*d*)ii, which has 24 copies of the core sequence. Differences in the lengths of fragments shown may be detected using Southern blotting and a suitable probe.

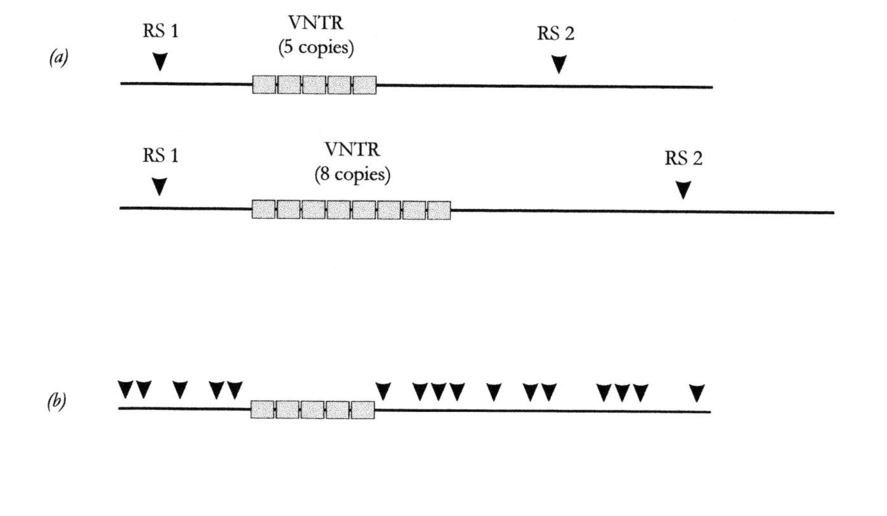

Fig. 9.8. Variable number tandem repeats (VNTRs) can generate what are essentially RFLPs, as shown in Fig. 9.7(*d*) and in this figure in (*a*). Here restriction sites 1 and 2 (RS1 and RS2) are separated by 5 and 8 copies of a repeat element for illustration. If fragments are produced by cutting at the recognition sites for RS 1 and RS2, different fragment lengths will be produced. An alternative method of analysing VNTRs is shown in (*b*). If a restriction enzyme is used that cuts the flanking DNA frequently (shown by arrowheads), but does NOT cut within the VNTR, the VNTR sequence is effectively isolated and trimmed, and left as a marker that can be used in DNA profiling (fingerprinting).

repeats of short (10–100 base-pair) sequences. The number of elements in a minisatellite region can vary, and thus these are also known as **variable number tandem repeats** (**VNTRs**, Fig. 9.8). These can be used in mapping studies, and also formed the basis of **genetic fingerprinting** (discussed in Chapter 11). One drawback with VNTRs is that they are not evenly distributed in the genome, tending to be located at the ends of chromosomes. Microsatellites have been used to overcome this difficulty. These are much shorter repeats, and the two base-pair CA repeat has been used as the standard microsatellite in mapping studies. Regions of CA repeats can be amplified using PCR, with primers that flank the repeated elements. As the primers are derived from unique-sequence regions, this essentially means that microsatellites amplified in this way are a type of STS (see Table 9.1), and can therefore be used to link genetic and physical maps.

Development of physical mapping techniques along with genetic mapping has enabled more detailed analysis of the genome in the latter half of the 1990s. A physical map with over 15000 STS markers was published in 1995, and by 1998 this had been extended to 30000 physical markers. Along with this increasing level of coverage and resolution, various types of software were developed to enable the map data to be correlated and viewed sensibly, and to be integrated with the emerging sequence data. The end result of all this activity was that, in a relatively short period, mapping of the genome had been moved on much more than many people had ever dared to hope.

9.4.3 Deriving and assembling the sequence

Sequence determination is perhaps the least troublesome area of genome analysis – particularly now that automated sequencing is the norm for large projects. However, it is not quite so simple! Although the technical aspects of sequencing are now well established, collating and annotating the data presents an enormous challenge (dealing with 3 billion As, Gs, Ts and Cs is not trivial). As with physical mapping, progress has been more rapid than many thought possible, and reliable methods for generating, collating and analysing DNA sequences have now been established.

Sequencing the genome has been carried out using a number of different techniques, although all have similar aims. The key point is the production of what are called **sequence-ready clones**. In theory, obtaining the human genome sequence could be done using a straightforward **shotgun** approach. However, the limiting step in this method is the task of putting the sequences together; this is essentially impossible for a genome as large as the human genome, in which there are many repetitive sequences. A **directed shotgun** approach, in which there is some attempt to link the shotgun sequences to map data, is better but is still likely to generate some anomalies.

The favoured method for large-scale sequencing is based on the **clone contig** method. In this, sequences from one clone are used to identify similar sequences in contiguous genome regions from other clones, either by hybridisation or by PCR amplification. Clones with minimum (but unambiguous) overlap can be selected for further processing. This method is the one most favoured to produce accurate sequence in finished form, as it links clones together and can therefore be checked easily by multiple sequencing of different clones where this is required. The use of BAC vectors is a popular and efficient method of generating the DNA fragments; some 300000 of these could, in theory, represent the entire human genome. By determining the sequences of the ends of the clones (to assist with clone ordering), and then assembling each clone sequence by a small-scale shotgun approach, sequence data can be generated in a controlled and accurate way.

The sequence information itself is, in most cases, placed in the public domain shortly after it has been processed. These databases are easily accessible on the World Wide Web. Anyone sitting at a personal computer can access the information for many different sequencing projects in addition to the human genome, and research in the virtual world is now a reality as people search sequence databases for interesting genes or control sequences.

9.4.4 What next?

The emphasis in genome research is now not so much on what the sequence *is*, but on what it *does*. Over the next few years, interpretation of the sequence will begin to integrate the fields of genomics and proteomics, and we will begin to get a greater understanding of how cells and tissues work at the molecular level. So what sort of impact will this have?

Undoubtedly a major focus will be on the medical advances that will arise as a result of genome research. The wealth of information that already exists for genetically based diseases (see **Online Mendelian Inheritance in Man**, as noted above) will be supplemented with more detailed analysis at the molecular level, and this will facilitate new treatments in areas such as drug design and gene therapy (discussed more fully in Chapter 11).

A development of mapping that adds a further level of detail to genetic and physical maps is the identification of **single nucleotide polymorphisms (SNPs)**. These are, as the name suggests, regions in which a single base is different between individuals. Thus one sequence might read AGTT**C**GATGCG, and in another person might be AGTT**A**GATGCG, with the C at position 5 changed to an A. SNPs are (by definition) so small that they have not been subjected to the usual pressures of evolution, as most of these changes will not affect the reproductive fitness of an individual. Thus they have become scattered evenly across the genome, one every 1300–1500 base-pairs or so. Whilst many will perhaps not be directly attributable as the cause of a particular disease, they will be used as useful markers to bring an increased level of subtlety to genome-based diagnosis.

The use of 'DNA chips' will also move the analysis of genomes on considerably. The chip consists of a microarray of short DNA sequences attached to a support, with each of the many sequences (in theory up to a million per cm^2) available for hybridisation to a sample DNA or RNA sequence. The technology has developed rapidly, and can now be used to screen for the expression of certain genes by using arrays of DNA sequences to probe mixtures of mRNAs. It can also be used to identify SNPs, and for novel methods of DNA sequencing.

Despite all the positive developments, concerns about the misuse of genome information have been around for as long as the project has been under way. There are very real **ethical** concerns. These range from the dilemma of whether or not to tell someone of a latent genetic condition that will appear in later life, to the use of genome information to discriminate against someone in areas such as life assurance or career path. There are also **legal** aspects that are difficult – the patenting of gene sequence information being one particularly troublesome area. However, these aspects are not peculiar to the genome project, as ethical dilemmas occur in many areas of science and technology. With common sense and a solid regulatory framework, many of the concerns can be alleviated. The whole aspect of ethics in genome analysis and gene manipulation generally is sometimes called **ELSI** (**E**thical, **L**egal and **S**ocial **I**mplications) and will be discussed further in Chapter 14.

The rate of progress in the human genome project has been staggering. Determination of the finished form of the sequence itself, once the 'holy grail' of genetics, is now within reach. On 15th February 2001, the journal *Nature* published its genome issue – in the context of free access to genome information from publicly funded research, this was available online at URL [**http://www.nature.com**]. This piece of text was written on 16th February 2001; as I write, I am wondering what lies ahead for genome research in the months until the publication of this book. By the time you read this, we will undoubtedly have moved on to yet another level of understanding of how our genome makes us what we are.

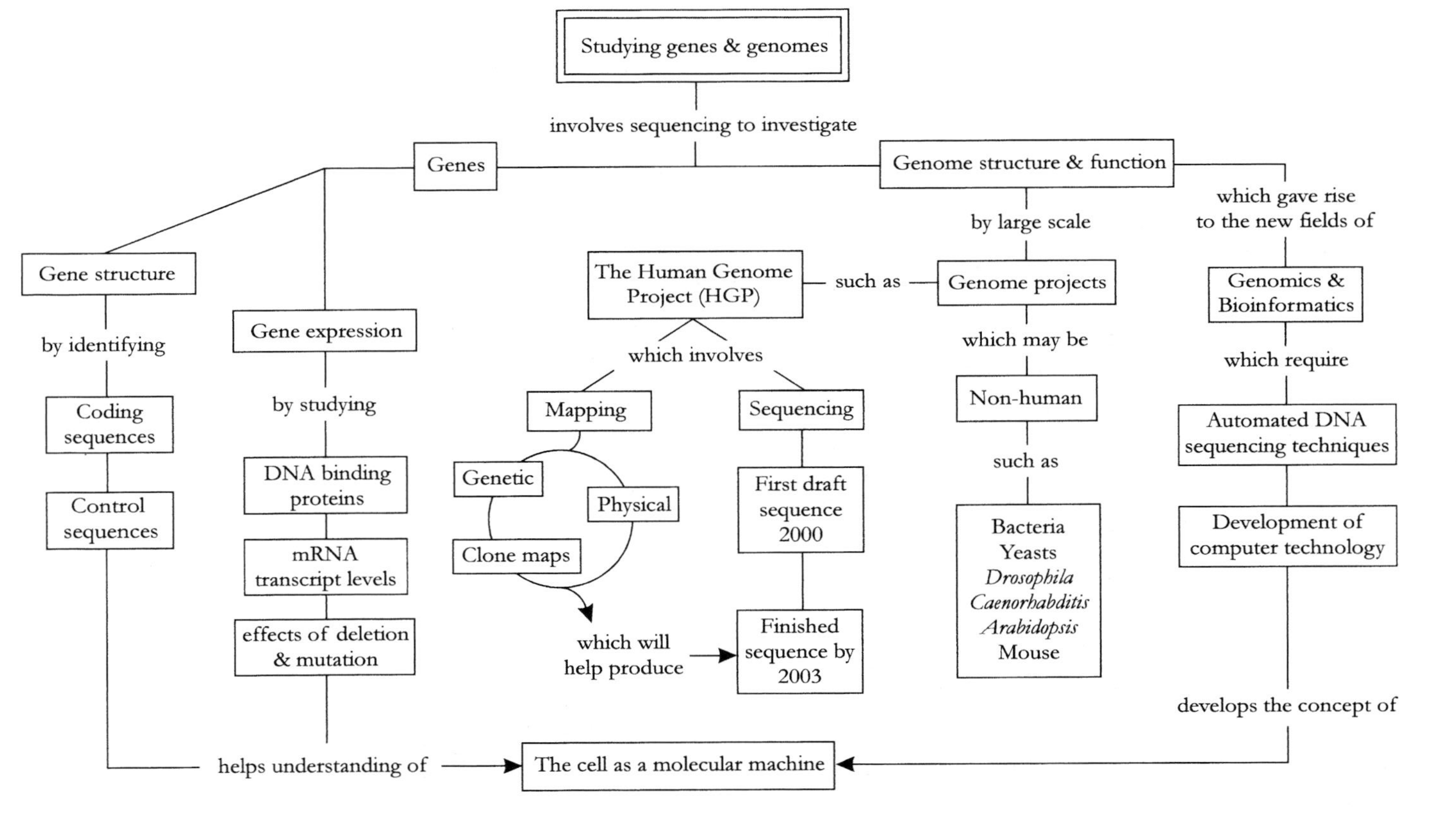

Concept map 9

10

Genetic engineering and biotechnology

Biotechnology is one of those difficult terms that can mean different things to different people. In essence, it is the use of an organism (usually a micro-organism) or a biologically derived substance (usually an enzyme) in a production or conversion process. Thus brewing and wine-making, food processing and manufacture, the production of pharmaceuticals and even the treatment of sewage can all be classed as aspects of biotechnology. In many cases the organism or enzyme is used in its natural form, and is not modified apart from perhaps having been subjected to selection methods to enable the best strain or type of enzyme to be used for a particular application. However, despite its traditional roots, modern biotechnology is often associated with the use of genetically modified systems. In this chapter we will consider the impact that gene manipulation technology has had on some biotechnological applications, with particular reference to the production of useful proteins.

The products of biotechnological processes are destined for use in a variety of fields such as medicine, agriculture and scientific research. It is perhaps an arbitrary distinction to separate the production of a therapeutic protein from its clinical application, as both could be considered as 'biotechnology' in its broadest sense. In a similar way, the developing area of transgenic plants and animals is also part of biotechnology, and undoubtedly the information provided by genome sequencing will give rise to many more diverse biotechnological applications. The impact of gene manipulation in medicine and transgenics will be considered in more detail in Chapters 11 and 12.

The production of recombinant proteins is now a well-developed area of research and development, and there is a bewildering range of different vector/host combinations. Whilst many workers may still wish to develop their own vectors with specific characteristics, it is now possible to buy vector/host combinations that suit most common applications. As with basic cloning vectors, the commercial opportunities presented by increasing demand for sophisticated expression systems have been exploited by many suppliers, and a look through supplier's catalogues or websites is a good way to get an overview of the current state of the technology.

10.1 Making proteins

The synthesis and purification of proteins from cloned genes is one of the most important aspects of genetic manipulation, particularly where valuable therapeutic proteins are concerned. Many such proteins have already been produced by recombinant DNA techniques, and are already in widespread use. We will consider some examples later in this chapter. In many cases a bacterial host cell can be used for the expression of cloned genes, but often a eukaryotic host is required for particular purposes.

In protein production there are two aspects which require optimisation, these being: (i) the biology of the system and (ii) the production process itself. Careful design of both these aspects is required if the overall process is to be commercially viable, which is necessary if large-scale production and marketing of the protein is the aim. Thus biotechnological applications require both a biological input and a process engineering input if success is to be achieved, and one of the key challenges in designing a particular process is the scale-up from laboratory to production plant.

10.1.1 Native and fusion proteins

For efficient expression of cloned DNA, the gene must be inserted into a vector that has a suitable promoter (see Table 6.3), and which can be introduced into an appropriate host such as *E. coli*. Although this organism is not ideal for expressing eukaryotic genes, many of the problems of using *E. coli* can be overcome by constructing the recombinant so that the expression signals are recognised by the host cell. Such signals include promoters and terminators for transcription, and ribosome binding sites (Shine–Dalgarno sequences) for translation. Alternatively, a eukaryotic host such as the yeast *S. cerevisiae*, or mammalian cells in tissue culture, may be more suitable for certain proteins.

For eukaryotic proteins, the coding sequence is usually derived from a

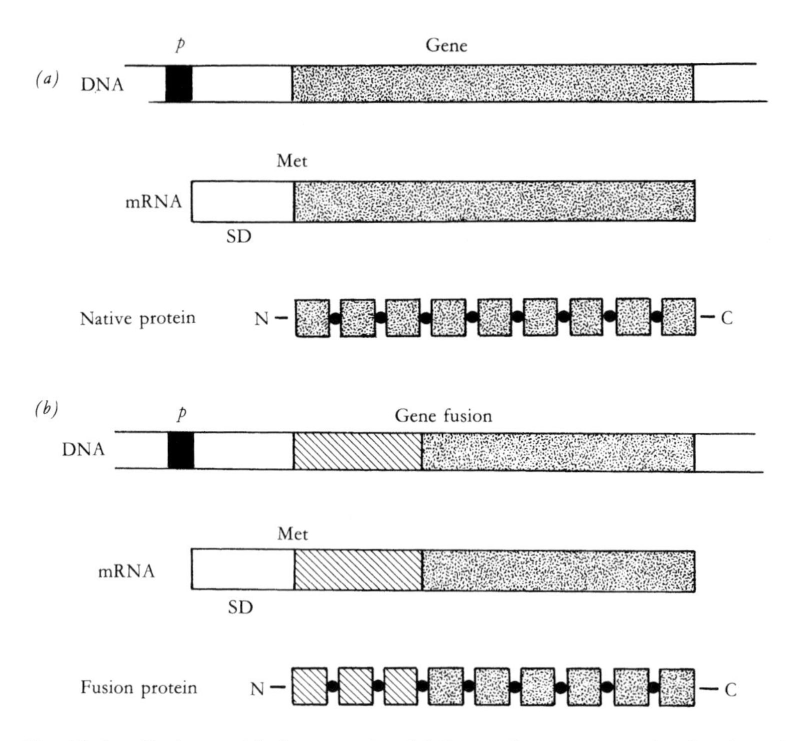

Fig. 10.1. Native and fusion proteins. (*a*) The coding sequence for the cloned gene (shaded) is not preceded by bacterial coding sequence, thus the mRNA encodes only insert-specified amino acid residues. This produces a native protein, synthesised from its own N terminus. (*b*) The gene fusion contains bacterial codons (hatched), therefore the protein contains part of the bacterial protein. In this example the first three N-terminal amino acid residues are of bacterial origin (hatched). The ribosome-binding site, or Shine–Dalgarno sequence, is marked SD.

cDNA clone of the mRNA. This is particularly important if the gene contains introns, as these will not be processed out of the primary transcript in a prokaryotic host. When the cDNA has been obtained, a suitable vector must be chosen. Although there is a wide variety of expression vectors, there are two main categories, which produce either **native proteins** or **fusion proteins** (Fig. 10.1). Native proteins are synthesised directly from the N terminus of the cDNA, whereas fusion proteins contain short, N-terminal amino acid sequences encoded by the vector. In some cases these may be important for protein stability or secretion, and are thus not necessarily a problem. However, such sequences can be removed if the recombinant is constructed so that the fusion protein contains a methionine residue at the point of fusion. The chemical cyanogen bromide (CNBr) can be used to cleave the protein at the

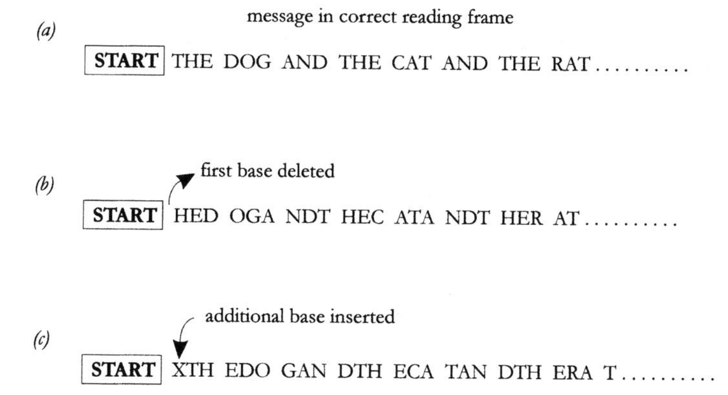

Fig. 10.2. The importance of reading frame. A simple sentence is used for illustration. In (*a*) the message has been 'cloned' downstream from the start site and is readable, as it is in the correct reading frame. In (*b*) a deletion of one base at the start is enough to knock out the sense completely. Addition of an extra base also causes problems as shown in (*c*).

methionine residue, thus releasing the desired peptide. A major problem with this approach occurs if the protein contains one or more internal methionine residues, as this will result in unwanted cleavage by CNBr.

When constructing a recombinant for the synthesis of a fusion protein, it is important that the cDNA sequence is inserted into the vector in a position that maintains the correct **reading frame.** The addition or deletion of one or two base-pairs at the vector/insert junction may be necessary to ensure this, although there are vectors that have been constructed so that all three potential reading frames are represented for a particular vector/insert combination. Thus by using the three variants of the vector, the correct in-frame fusion can be obtained. The importance of reading frame is shown in Fig. 10.2.

10.1.2 Yeast expression systems

As discussed in Chapter 5, the yeast *Saccharomyces cerevisiae* has been the favoured microbial eukaryote in the development of recombinant DNA technology. Other yeasts that are used include *Schizosaccharomyces pombe*, *Pichia pastoris*, *Hansela polymorpha*, *Kluyveromyces lactis*, and *Yarrowia lipolytica*. These demonstrate many of the characteristics of bacteria with respect to ease of use – they grow rapidly on relatively inexpensive media, and a range of different mutant strains and vectors is available for various applications. In some cases scale-up fermentations present some difficulties compared to bacteria, but these can usually be overcome by careful design and monitoring of the process. Yields of heterologous proteins of around $12\,g\,l^{-1}$ (10–100 times more than in *S. cerevisiae*) have been obtained using *P. pastoris*, which can be grown on methanol as sole carbon source. In this situation growth is regulated by alcohol oxidase, which has a low specific activity and is consequently overproduced in these cells, making up around 30% of total soluble protein. By placing heterologous genes downstream from the alcohol oxidase promoter (AOX1), high levels of expression are achieved.

One of the advantages of using yeast as opposed to bacterial hosts is that proteins are subjected to post-translational modifications such as glycosylation. In addition, there is usually a higher degree of 'authenticity' with respect to 3D conformation and the immunogenic properties of the protein. Thus, in a situation where the biological properties of the protein are critical, yeasts may provide a better product than prokaryotic hosts.

10.1.3 The baculovirus expression system

Baculoviruses infect insects, and do not appear to infect mammalian cells. Thus any system based on such viruses has the immediate attraction of low risk of human infection. During normal infection of insect cells, virus particles are packaged within **polyhedra**, which are nuclear inclusion bodies composed mostly of the protein **polyhedrin**. This is synthesised late in the virus infection cycle, and can represent as much as 50% of infected cell protein when fully expressed. Whilst polyhedra are required for infection of insects themselves, they are not required to maintain infection of cultured cells. Thus the polyhedrin gene is an obvious candidate for construction of an expression vector, as it encodes a late-expressed dispensable protein that is synthesised in large amounts.

The baculovirus genome is a circular double-stranded DNA molecule. Genome size is from 88 to 200 kb, depending on the particular virus, and the genome is therefore too large to be manipulated directly. Thus insertion of foreign DNA into the vector has to be accomplished by using an intermediate known as a **transfer vector**. These are based on *E. coli* plasmids, and carry the promoter for the polyhedrin gene (or for another viral gene) and any other essential expression signals. The cloned gene for expression is inserted into the transfer vector, and the recombinant is used to co-transfect insect cells with non-recombinant viral DNA. Homologous recombination between the viral DNA and the transfer vector results in the generation of recombinant viral genomes, which can be selected for and used to produce the protein of interest.

10.1.4 Mammalian cell lines

Where the expression of recombinant human proteins is concerned, it might seem obvious that a mammalian host cell would be a better system that bacteria or even eukaryotic microbes. However, the use of such cell lines in protein production presents some problems. Often media for growth are complex and expensive, and mammalian cells are generally less robust that microbes when large-scale fermentation is involved. There may also be difficulties in the processing of the products (often the term **downstream processing** is used to describe the operations needed to purify a protein from a fermentation process). Despite these difficulties, many vectors are now available for protein expression in mammalian cells. They exhibit characteristics that will by now be familiar – often based on a viral system, vectors utilise selectable markers (often drug-resistance markers) and have promoters that enable expression of the cloned gene sequence. Common promoters are based on Simian Virus (SV40) or Cytomegalovirus (CMV). Some examples were presented in Table 5.3.

10.2 Protein engineering

One of the most exciting applications of gene manipulation lies in the field of protein engineering. This involves altering the structure of proteins via alterations to the gene sequence, and has become possible due to the technique of mutagenesis *in vitro*. In addition, a deeper understanding of the structural and functional characteristics of proteins has enabled workers to pinpoint the essential amino acid residues in a protein sequence. Alterations can therefore be carried out at these positions and their effects studied. The desired effect might be alteration of the catalytic activity of an enzyme by modification of the residues around the active site, an improvement in the nutritional status of a storage protein, or an improvement in the stability of a protein used in industry or medicine.

Mutagenesis *in vitro* enables specific mutations to be introduced into a gene sequence. One technique is called **oligonucleotide-directed** or **site-directed** mutagenesis, and is elegantly simple in concept (Fig. 10.3). The

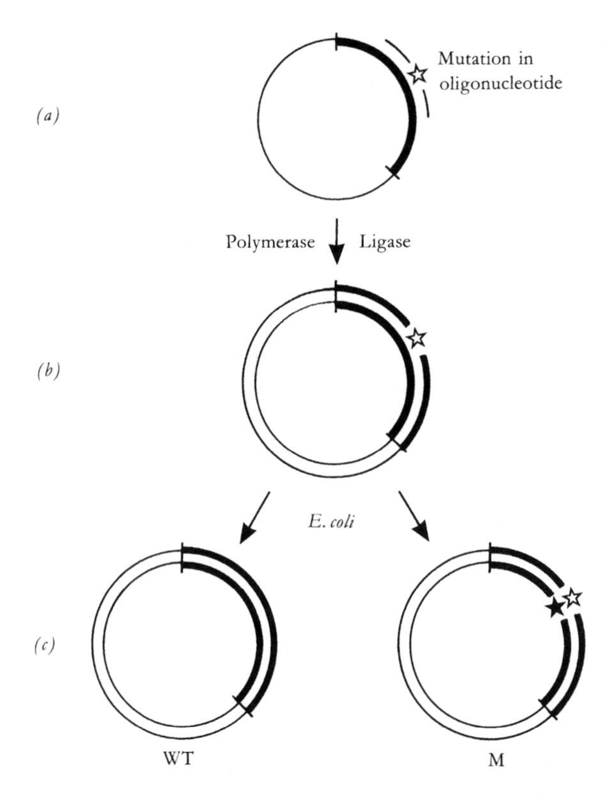

Fig. 10.3. Oligonucleotide-directed mutagenesis. (*a*) The requirement for mutagenesis *in vitro* is a single-stranded DNA template containing a cloned gene (heavy line). An oligonucleotide is synthesised that is complementary to the part of the gene that is to be mutated (but which incorporates the desired mutation). This is annealed to the template (the mutation is shown as an open star). (*b*) The molecule is made double-stranded in a reaction using DNA polymerase and ligase, which produces a hybrid wild-type/mutant DNA molecule with a mismatch in the mutated region. (*c*) On introduction into *E. coli* the molceule is replicated, thus producing double-stranded copies of the wild-type (WT) and mutant (M) forms. The mutant carries the original mutation and its complementary base or sequence (filled star).

requirements are a single-stranded template containing the gene to be altered, and an oligonucleotide (usually 15–30 nucleotides in length) that is complementary to the region of interest. The oligonucleotide is synthesised with the desired mutation as part of the sequence. The single-stranded template is often produced using the M13 cloning system, which produces ss DNA. The template and oligonucleotide are annealed (the mutation site will mismatch, but the flanking sequences will confer stability), and the template is then copied using DNA polymerase. This gives rise to a ds DNA which, on replication, will yield two daughter molecules, one of which will contain the desired mutation.

Identification of the mutated DNA can be carried out by hybridisation with the mutating oligonucleotide sequence, which is radiolabelled. Non-mutated DNA will retain the original mismatch, whereas the mutant will match perfectly. By washing filters of the suspected mutants at high stringency, all imperfect matches can be removed and the mutants detected by autoradiography. Even a single base-pair change can be picked up using this technique. The mutant can then be sequenced to confirm its identity.

Having altered a gene by mutagenesis, the protein is produced using an expression system. Often a vector incorporating the *lac* promoter is used, so that transcription can be controlled by the addition of IPTG. Alternatively, the λP_L promoter can be used with a temperature-sensitive λcI repressor, so that expression of the mutant gene is repressed at 30 °C but is permitted at 42 °C. Analysis of the mutant protein is carried out by comparison with the wild-type protein. In this way, proteins can be 'engineered' by incorporating subtle structural changes that alter their functional characteristics. This technique has great potential that is only just beginning to be exploited.

10.3 Examples of biotechnological applications of rDNA technology

In the final part of this chapter we will consider some examples of the types of product that can be produced using recombinant DNA technology in biotechnological processes. This is a rapidly developing area for many biotechnology companies, with large-scale investment in both basic research and in development to production status. This aspect of gene manipulation technology is likely to become increasingly important in the future, particularly in medicine and general healthcare, with many diverse products being brought to market.

10.3.1 Production of enzymes

The commercial production and use of enzymes is already a well-established part of the biotechnology industry. Enzymes are used in brewing, food processing, textile manufacture, the leather industry, washing powders, medical applications, and basic scientific research, to name just a few examples. In many cases the enzymes are prepared from natural sources, but in recent years there has been a move towards the use of enzymes produced by recombinant DNA methods, where this is possible. In addition to the scientific problems of producing a recombinant-derived enzyme, there are economic factors to take into account, and in many cases the cost–benefit analysis makes the use of a recombinant enzyme unattractive. Broadly speaking, enzymes are either high-volume/low cost preparations for use in industrial scale operations, or are low-volume/high value products that may have a very specific and relatively limited market.

There is a nice twist to the gene manipulation story in that some of the enzymes used in the procedures are now themselves produced using rDNA methods. Many of the commercial suppliers list recombinant variants of the common enzymes, such as polymerases (particularly for PCR) and others. Recombinant enzymes can sometimes be engineered so that their characteristics fit the criteria for a particular process better than the natural enzyme, which increases the fidelity and efficiency of the process.

In the food industry, one area that has involved the use of recombinant enzyme is the production of cheese. In cheese manufacture, **rennet** (also known as rennin, chymase, or **chymosin**) has been used as part of the process. Chymosin is a protease that is involved in the coagulation of milk casein following fermentation by lactic acid bacteria. It was traditionally prepared from animal (bovine or pig) or fungal sources. In the 1960s the Food and Agriculture Organisation of the United Nations predicted that a shortage of calf rennet would develop as more calves were reared to maturity to satisfy increasing demands for meat and meat products. Today, there are six sources for natural chymosin – veal calves, adult cows and pigs, and the fungi *Rhizomucor miehei*, *Endothia parasitica* and *Rhizomucor pusillus*. Chymosin is now also available as a recombinant-derived preparation from *E. coli*, *Kluyveromyces lactis* and *Aspergillus niger*. Recombinant chymosin was first developed in 1981 and approved in 1988, and is now used to prepare around 90% of hard cheeses in the UK.

Although the public acceptance of what is loosely called 'GM cheese' has not presented as many problems as has been the case with other areas of gene manipulation of foodstuffs, there are still concerns that need to be addressed. In cheese manufacture, there are three possible objections that can be raised by those who are concerned about GM foods. Firstly, milk could have been produced from cows treated with recombinant growth hormone (see Section 10.3.2 below). Secondly, they could have been fed with animal feeds containing GM soya or maize. The third concern is the use of recombinant-derived chymosin. Despite these fears, many consumers are content that cheese is not itself a **Genetically Modified Organism** (**GMO**), but is the *product* of a *product* of a GMO.

A final example of recombinant-derived proteins in consumer products is the use of enzymes in washing powder. **Proteases** and **lipases** are commonly used to assist cleaning by degradation of protein and lipid-based staining. A recombinant lipase was developed in 1988 by Novo Nordisk A/V (now known as Novozymes). The company is the largest supplier of enzymes for commercial use in cleaning applications. Their recombinant lipase was known as **Lipolase**, which was the first commercial enzyme developed using rDNA technology and the first lipase used in detergents. A further development involved an engineered variant of Lipolase called Lipolase Ultra, which gives enhanced fat removal at low wash temperatures.

10.3.2 The BST story

Not all rDNA biotechnology projects have a smooth passage from inception to commercial success. The development of recombinant **bovine somatotropin** (**rBST**) illustrates some of the problems that may be encountered once the scientific part of the process has been achieved. In bringing a recombinant product such as rBST (and the examples outlined above) to market, many aspects have to be considered. The **basic science** has to be carried out, followed by **technology transfer** to get the process to a commercially viable stage. **Approval** by regulatory bodies may be required, and finally (and most critical from a commercial standpoint) the product has to gain **market acceptance** and establish a consumer base. We can find all of these aspects in the BST story.

BST is also known as bovine growth hormone, and is a naturally occurring protein that acts as a growth promoter in cattle. Milk production can be increased substantially by administering BST, and thus it was an attractive target for cloning and production for use in the dairy industry. The basic science of rBST was relatively straightforward, and scientists were already working on this in the early 1980s. The BST gene was in fact one of the first mammalian genes to be cloned and expressed, using bacterial cells for production of the protein. Thus the production of rBST at a commercial level, involving the basic science and technology transfer stages, was achieved without too much difficulty. A summary of the process is shown in Fig. 10.4.

With respect to approval of new rDNA products, each country has its own system. In the USA, the **Food and Drug Administration** (**FDA**) is the central regulatory body, and in 1994 approval was given for the commercial

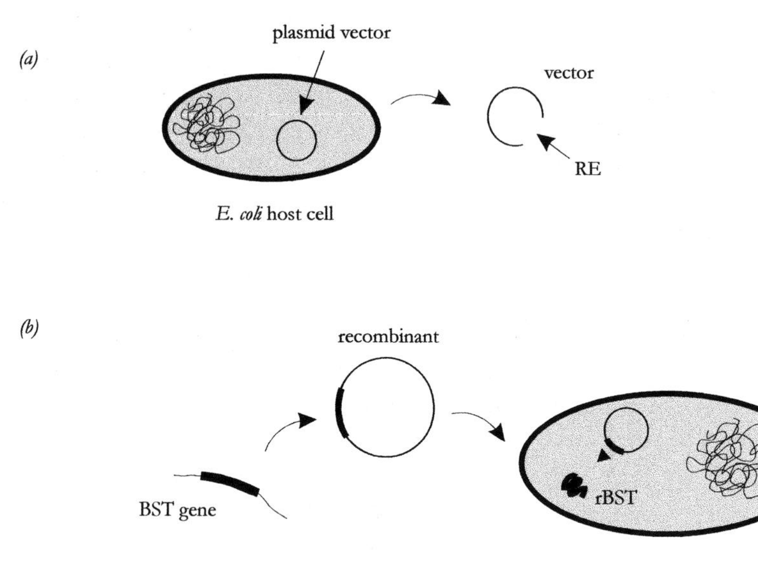

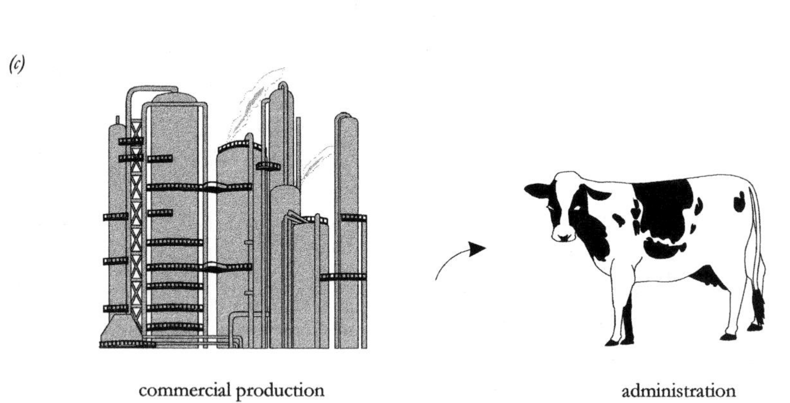

Fig. 10.4. Production of recombinant bovine growth hormone (rBST). (*a*) A plasmid vector is prepared from *E. coli* and cut with a restriction enzyme (RE). (*b*) The BST gene coding sequence is ligated into the plasmid to generate the recombinant, which produces rBST protein in the cell following transformation. Scale-up to commercial production is shown in (*c*), and with product approval granted, administration can begin. The whole process from basic science to market usually takes several/many years from start to finish, with a large amount of investment capital required. From Nicholl (2000), *Cell and Molecular Biology*, Advanced Higher Monograph Series, Learning and Teaching Scotland. Reproduced with permission.

distribution of rBST, marketed by Monsanto under the trade name **Posilac**. At that time the European Union did not approve the product, but this was partly for socioeconomic reasons (increasing milk production was not necessary) rather than for any concerns about the science. Evaluation of evidence at that time suggested that milk from rBST-treated cows was identical to normal untreated milk, and it was therefore unlikely that any negative effects would be seen in consumers.

The effects of rBST must be considered in three different contexts – the effect on milk production, the effects on the animals themselves, and the possible effects on the consumer. Milk production is usually increased by around 10–15% in treated cows, although yield increases of much more than this have been reported. Thus, from a dairy herd management viewpoint, use of rBST would seem to be beneficial. However, as is usually the case with any new development that is aimed at 'improving' what we eat or drink, public concern grew along with the technology. The concerns fuel a debate that is still ongoing, and is at times emotive. One area that is hotly debated is the effect of rBST on the cows themselves. Administering rBST can produce localised swelling at the site of injection, and can exacerbate problems with foot infections, mastitis and reproduction. The counter-argument is that many of these problems occur anyway, even in herds that are rBST-free. On balance, the evidence does, however, suggest that animal welfare is compromised to some extent when rBST is used.

The possible effect of rBST use on human health that is another area of great concern and debate. The natural hormone (and therefore the recombinant version also) affects milk production by increasing the levels of **insulin-like growth factor** (**IGF-1**), which causes increased milk production. Administration of rBST generates elevated levels of IGF-1, and there is evidence that IGF-1 can stimulate the growth of cancer cells. Thus the concern is that using rBST could pose a risk to health. The counter-argument in this case is that the levels of IGF-1 in the early stages of lactation are higher than those generated by the use of rBST in cows 100 days after lactation begins, which is often when it is administered. This arguably means that milk from early lactating cows should not be drunk at all if there are any concerns about IGF-1. Those who oppose the use of rBST point out that, unlike a therapeutic protein that would be used for a limited number of patients, milk is consumed by most people, and any inherent risk, no matter how small, is therefore unacceptable. On the basis of this uncertainty, many countries have banned the use of rBST, citing both the animal welfare issue and the potential risk to health as reasons. The arguments look set to continue into the future as commercial, animal welfare and human health interests clash.

Following the debate provides an interesting illustration of the problems surrounding the use of gene technology, and of the need for objective assessment of risks and the avoidance of emotive judgements.

10.3.3 Therapeutic products for use in human health-care

Although the production of recombinant-derived proteins for use in medical applications does raise some ethical concerns, there is little serious criticism aimed at this area of biotechnology. The reason is largely that therapeutic products and strategies are designed to alleviate suffering, or to improve the quality of life for those who have a treatable medical condition. In addition, the products are used under medical supervision, and there is a perception that the corporate interests that tend to be highlighted in the food debate have less of an impact in the diagnosis and treatment of disease. In fact, there is just as much competition and investment risk associated with the medical products field as is the case in agricultural applications; there does, however, seem to be less emotive debate in this area. It is therefore apparently much more acceptable to the public. In addition to the actual *treatment* of conditions, the area of **medical diagnostics** is a large and fast-growing sector of the biotechnology market, with recombinant DNA technology involved in many aspects of this.

Recombinant DNA products for use in medical therapy can be divided into three main categories. Firstly, protein products may be used for **replacement** or **supplementation** of human proteins that may be absent or ineffective in patients with a particular illness. Secondly, proteins can be used in **specific disease therapy**, to alleviate a disease state by intervention. Thirdly, the production of **recombinant vaccines** is an area that is developing rapidly and which offers great promise. Some examples of therapeutic proteins produced using recombinant DNA technology are listed in Table 10.1. We will consider examples from each of the three areas outlined above to illustrate the type of approach taken in developing a therapeutic protein.

The widespread condition **diabetes mellitus** (**DM**) is usually caused by β-cells in the islets of Langerhans in the pancreas failing to produce adequate amounts of the hormone **insulin**. Many millions of people worldwide are affected by DM, and the World Health Organization estimates that the global incidence will double by 2025. Sufferers are classed as having either **insulin-dependent DM** (**IDDM**) or **non-insulin dependent DM** (**NIDDB**). Insulin-dependent patients obviously require the hormone, but many NIDDB patients also use insulin for satisfactory control of their condition.

Table 10.1. *Selected rDNA-derived therapeutic products for use in humans*

Product	Type	Trade name	FDA App.	Company	Use
Insulin	R/S	Humulin	1982	Eli Lilly	Diabetes treatment
Growth hormone	R/S	Protropin	1985	Genentech	Growth hormone deficiency in children
α-Interferon	SDT	Intron A	1986	Schering-Plough	Hairy cell leukaemia
Hepatitis B vaccine	V	Recombivax HB	1986	Merck & Co.	Hepatitis B prevention
Tissue plasminogen activator	SDT	Activase	1987	Genentech	Myocardial infarction
Growth hormone	R/S	Humatrope	1987	Eli Lilly	Growth hormone deficiency in children
Hepatitis B vaccine	V	Engerix-B	1989	SmithKline Beecham	Hepatitis B prevention
Factor VIII	R/S	Recombinate rAHF (anti-haemophiliac factor)	1992	Baxter Healthcare	Treatment of haemophilia
Factor VIII	R/S	Kogenate	1993	Bayer	Treatment of haemophilia
DNase	SDT	Pulmozyme	1993	Genentech	Treatment of cystic fibrosis symptoms

Note: Type refers to replacement/supplementation (R/S), specific disease therapy (SDT) or vaccine (V). Dates in column 4 refer to year of first approval by the FDA for use in the USA. Subsequent approvals for additional uses are not shown. Products are listed chronologically with respect to year of first approval. Trade names are registered trademarks of the companies involved; company names are as given in the approval, and may have changed due to corporate policy or merger, etc. Further information can be found on the FDA website at URL [http://www.fda.gov].

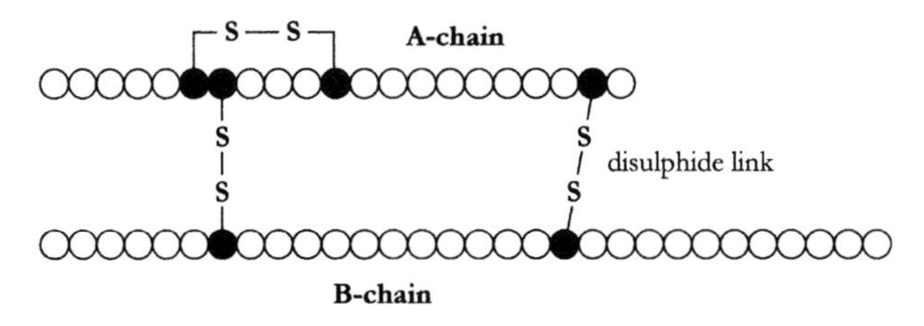

Fig. 10.5. The structure of insulin. Amino acids are represented by circles. The A-chain (21 amino acids) and B chain (30 amino acids) are held together by disulphide linkages between cysteine residues (filled circles).

Delivery of insulin is by injection, although novel methods of delivery such as inhalation of insulin powder are currently being developed.

Insulin is composed of two amino acid chains, the **A-chain** (acidic, 21 amino acids) and **B-chain** (basic, 30 amino acids). When synthesised naturally, these chains are linked by a further 30 amino acid peptide called the **C-chain.** This 81 amino acid precursor molecule is known as **proinsulin**. The A- and B-chains are linked together by disulphide bonds between cysteine residues, and the proinsulin is cleaved by a protease to produce the active hormone shown in Fig. 10.5. Insulin was the first protein to be sequenced, by Frederick Sanger in the mid-1950s.

As DM is caused by a problem with a normal body constituent (insulin), therapy falls into the category of **replacement** or **supplementation**. Banting and Best developed the use of insulin therapy in 1921, and for the next 60 or so years diabetics were dependent on natural sources of insulin, with the attendant problems of supply and quality. In the late 1970s and early 1980s recombinant DNA technology enabled scientists to synthesise insulin in bacteria, with the first approvals granted by 1982. Recombinant-derived insulin is now available in several forms, and has a major impact on diabetes therapy. One of the most widely used forms is marketed under the name **Humulin** by the Eli Lilly company.

In an early method for the production of recombinant insulin, the insulin A- and B-chains were synthesised separately in two bacterial strains. The insulin A- and B-genes were placed under the control of the *lac* promoter, so that expression of the cloned genes could be switched on by using lactose as the inducer. Following purification of the A- and B-chains, they were linked together by a chemical process to produce the final insulin molecule. The process is shown in Fig. 10.6. A development of this method involves the synthesis of the entire proinsulin polypeptide (shown in Fig. 10.7) from a single gene sequence. The product is converted to insulin enzymatically.

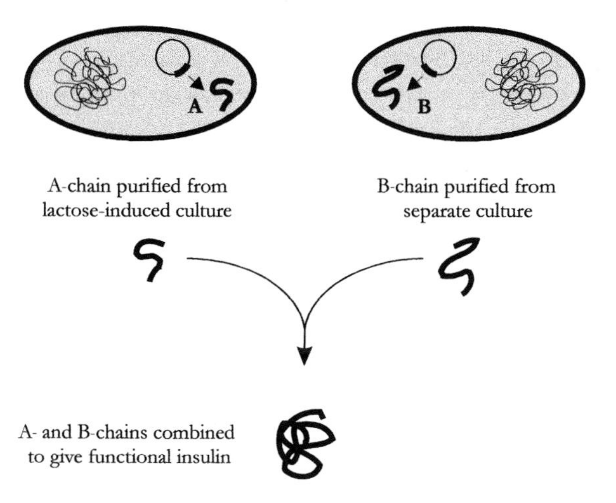

Fig. 10.6. Production of recombinant-derived insulin by separate fermentations for A- and B-chains. Lactose is used to induce transcription of the cloned gene sequences from the *lac* promoter. Following translation, the products are purified to give A- and B-chains that are then combined chemically to give the final product.

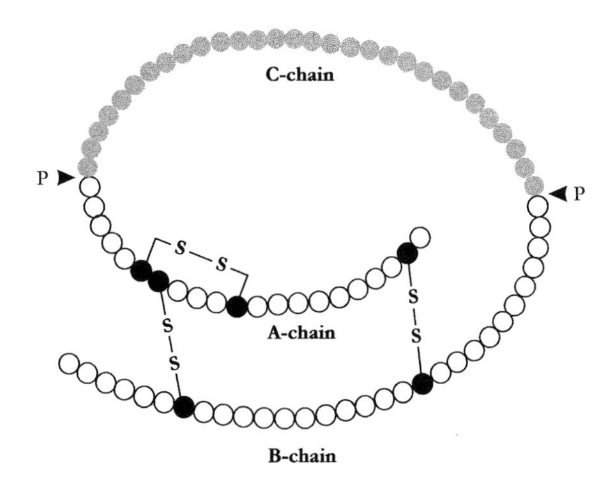

Fig. 10.7. Proinsulin. This molecular precursor of insulin is synthesised as an 81 amino acid polypeptide. The C peptide sequence is then removed by a protease (P) to leave the A- and B-chains to form the final insulin molecule. Proinsulin is now synthesised intact during rDNA-based production of insulin.

There are many recombinant proteins for use in **specific disease therapy**. One example of this type of protein is **tissue plasminogen activator (TPA)**. This is a protease that occurs naturally, and functions in breaking down blood clots. TPA acts on an inactive precursor protease called **plasminogen**, which is converted to the active form called **plasmin**. This protease attacks the clot by breaking up **fibrin**, the protein that is involved in clot formation. TPA is used as a treatment for heart attack victims. If administered soon after an attack, it can help reduce the damage caused by coronary thrombosis.

Recombinant TPA was produced in the early 1980s by the company Genentech using cDNA technology. It was licensed in the USA in 1987, under the trade name **Activase**, for use in treatment of acute myocardial infarction. It was the first recombinant-derived therapeutic protein to be produced from cultured mammalian cells, which secrete rTPA when grown under appropriate conditions. The amount of rTPA produced in this way was sufficient for therapeutic use, thus a major advance in coronary care was achieved. Further uses were approved in 1990 (for acute massive pulmonary embolism) and 1996 (for acute ischaemic stroke).

The final group of recombinant-derived products are **vaccines**. There are now many vaccines available for animals, and the development of human vaccines is also beginning to have an impact in healthcare programmes. One vaccine that has been produced by rDNA methods is the **hepatitis B** vaccine. The yeast *Saccharomyces cerevisiae* is used to express the surface antigen of the hepatitis B virus (**HBsAg**), under the control of the alcohol dehydrogenase promoter. The protein can then be purified from the fermentation culture, and used for inoculation. This removes the possibility of contamination of the vaccine by blood-borne viruses or toxins, which is a risk if natural sources are used for vaccine production.

A further development in vaccine technology involves using **transgenic plants** to deliver non-plant antigens to the patient. This area of research and development has tremendous potential, particularly for vaccine delivery in underdeveloped countries where traditional methods of vaccination may be not fully effective due to cost and distribution problems. The attraction of having a vaccine-containing banana or tomato is clear, and development and trials are currently under way for a variety of plant vaccines.

The use of gene manipulation techniques in the biotechnology industry is a major developing area of applied science. In addition to the scientific and engineering aspects of the work, the financing of biotechnology companies is an area that presents its own risks and potential rewards – for example, a new drug may take 10–15 years to develop, at a cost of around £350 million.

The stakes are therefore high, and many fledgling companies fail to survive their first few years of operation. Even established and well-financed companies are not immune to the risk associated with the development of a new and untried product. The next few years will certainly be interesting for this sector of the applied science industry.

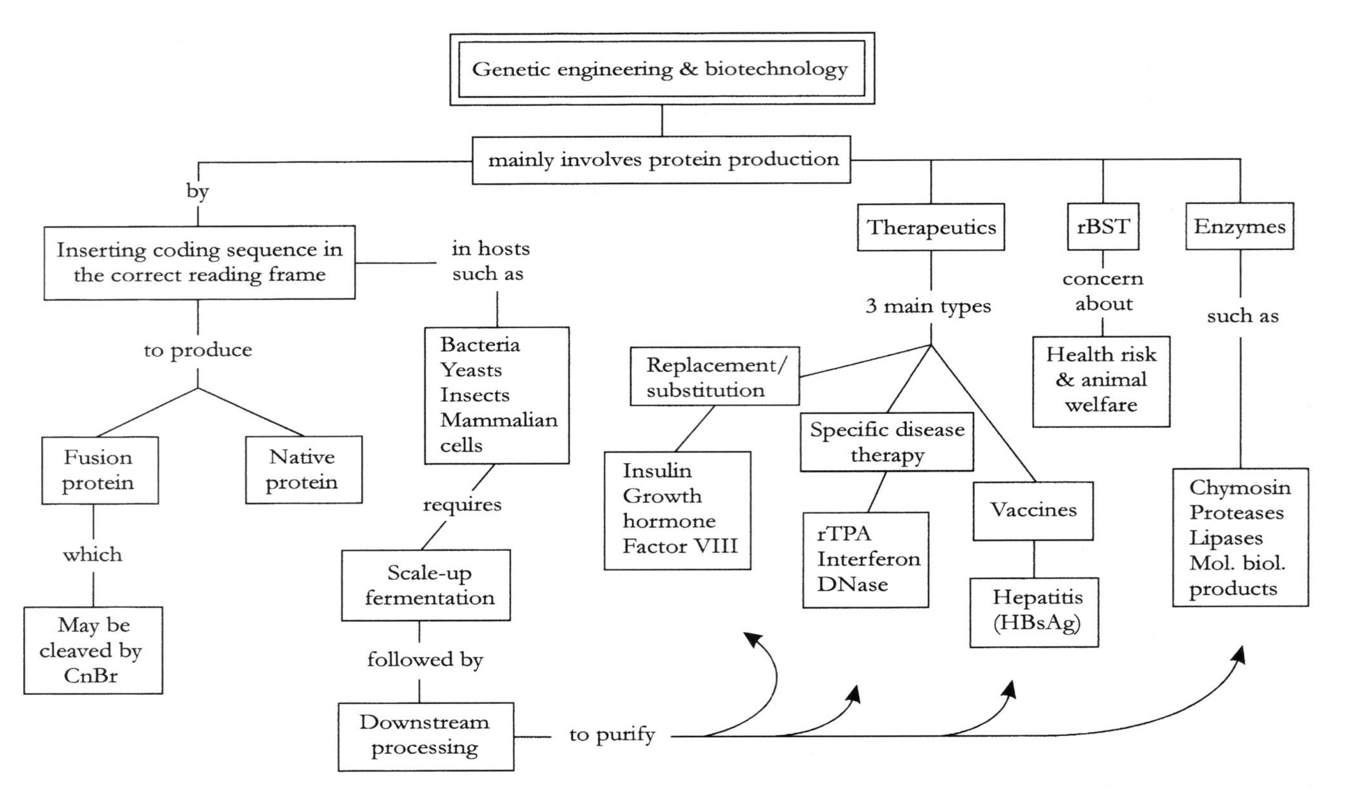

Concept map 10

11

Medical and forensic applications of gene manipulation

The diagnosis and treatment of human disease is one area in which genetic manipulation is beginning to have a considerable effect. As outlined in Chapter 10, many therapeutic proteins are now made by recombinant DNA methods, and the number available is increasing steadily. Thus the treatment of conditions by recombinant-derived products is already well established. In this chapter we will look at how the techniques of gene manipulation impact more directly on medical diagnosis and treatment, and will also examine the use of rDNA technology in forensic science. Progress in both of these areas is of course closely linked to our increasing knowledge of the human genome, and thus many new developments in medical and forensic applications will appear as we decipher the genome.

11.1 Diagnosis and characterisation of medical conditions

Genetically based diseases (often called simply 'genetic diseases') represent one of the most important classes of disease, particularly in children. A disorder present at birth is termed a **congenital** abnormality, and around 5% of newborn babies will suffer from a serious medical problem of this type. In most of these cases there will be a significant genetic component in the **aetiology** (cause) of the disease state. It is estimated that about a third of primary admissions to paediatric hospitals are due to genetically based problems, whilst some 70% of cases presenting more than once are due to genetic defects. In addition to genetic problems appearing at birth or in childhood, it seems that a large proportion of diseases presenting in later life also have a genetic cause or predisposition. Thus medical genetics, in its traditional non-recombinant form, has already had a major impact on the diagnosis of disease and abnormality. The development of molecular genetics and rDNA technology has not only broadened the range of techniques available for diagnosis, but has also opened up the possibility of novel gene-based treatments for certain conditions.

11.1.1 Diagnosis of infection

In addition to genetic conditions that affect the individual, rDNA technology is also important in the diagnosis of certain types of infection. Normally, bacterial infection is relatively simple to diagnose, once it has taken hold. Thus the prescription of antibiotics may follow a simple investigation by a general practitioner. A more specific characterisation of the infectious agent may be carried out using microbiological culturing techniques, and this is often necessary when the infection does not respond well to treatment. Viral infections may be more difficult to diagnose, although conditions such as *Herpes* infections are usually obvious.

Despite traditional methods being applied in many cases, there may be times when these methods are not appropriate. Infection by the **human immunodeficiency virus** (**HIV**) is one case in point. The virus is the causative agent of **acquired immune deficiency syndrome** (**AIDS**). The standard test for HIV infection requires immunological detection of anti-HIV antibodies. However, these antibodies may not be detectable in an infected person until weeks after initial infection, by which time others may have been infected. A test such as this, where no positive result is obtained even though the individual is infected, is a **false negative**. The use of DNA probes and PCR technology circumvents this problem by assaying for the actual viral DNA in the T-lymphocytes of the patient, thus permitting a diagnosis before the antibodies are detectable.

Other examples of the use of rDNA technology in diagnosing infections include **tuberculosis** (caused by the bacterium *Mycobacterium tuberculosis*), **human papilloma virus** (**HPV**) infection, and **Lyme disease** (caused by the spirochaete *Borrelia burgdorferi*).

11.1.2 Patterns of inheritance

Although diagnosis of infection is an important use of rDNA technology, it is in the characterisation of genetic disease that the technology has perhaps been most applied in medicine to date. Before dealing with some specific diseases in more detail, it may be useful to review the basic features of transmission genetics, and outline the range of factors that may determine how a particular disease state presents in a patient.

Since it was rediscovered in 1900, the work of Gregor Mendel has formed the basis for our understanding of how genetic characteristics are passed on from one generation to the next. We have already seen that the human genome is made up of some 3 billion base-pairs of information. This is organised as a **diploid** set of 46 chromosomes, arranged as 22 pairs of **autosomes** and one pair of **sex chromosomes**. Prior to reproduction, the **haploid** male and female **gametes** (sperm and oocyte respectively) are formed by the reduction division of **meiosis**, which reduces the chromosome number to 23. On fertilisation of the oocyte by the sperm, diploid status is restored, with the **zygote** receiving one member of each chromosome pair from the father, and one from the mother. In males the sex chromosomes are **X** and **Y**, in females **XX**, and thus it is the father that determines the sex of the child.

Traits may be controlled by single genes, or by many genes acting in concert. Single-gene disease traits are known as **monogenic** disorders, whilst those involving many genes are **polygenic**. Inheritance of a monogenic disease trait usually follows a basic Mendelian pattern, and can therefore often be traced in family histories by pedigree analysis. A gene may have **alleles** (different forms) that may be **dominant** (exhibited when the allele is present) or **recessive** (the effect is masked by a dominant allele). With respect to a particular gene, individuals are said to be either **homozygous** (both alleles the same) or **heterozygous** (the alleles are different, perhaps one dominant and one recessive). Patterns of inheritance of monogenic traits can be associated with the autosomes, as either **autosomal dominant** or **autosomal recessive**, or may be sex-linked (usually with the X chromosome, thus showing **X-linked** inheritance). The Mendelian patterns and ratios for these types of inheritance are shown in Fig. 11.1. In addition to the nuclear chromosomes, mutated genes associated with the mitochondrial genome can cause disease. As the mitochondria are inherited along with the egg, these traits show **maternal** patterns of inheritance. We will consider specific examples of the patterns of inheritance outlined above in the next section.

The effect of a gene depends not only on its allelic form and character, but also on how it is expressed. The terms **penetrance** and **expressivity** are used to describe this aspect. Penetrance is usually quoted as the percentage of individuals carrying a particular allele who demonstrate the associated phenotype. Expressivity refers to the degree to which the associated phenotype is presented (the *severity* of the phenotype is one way to think of this). Thus alleles showing **incomplete penetrance** and/or **variable expressivity** can greatly

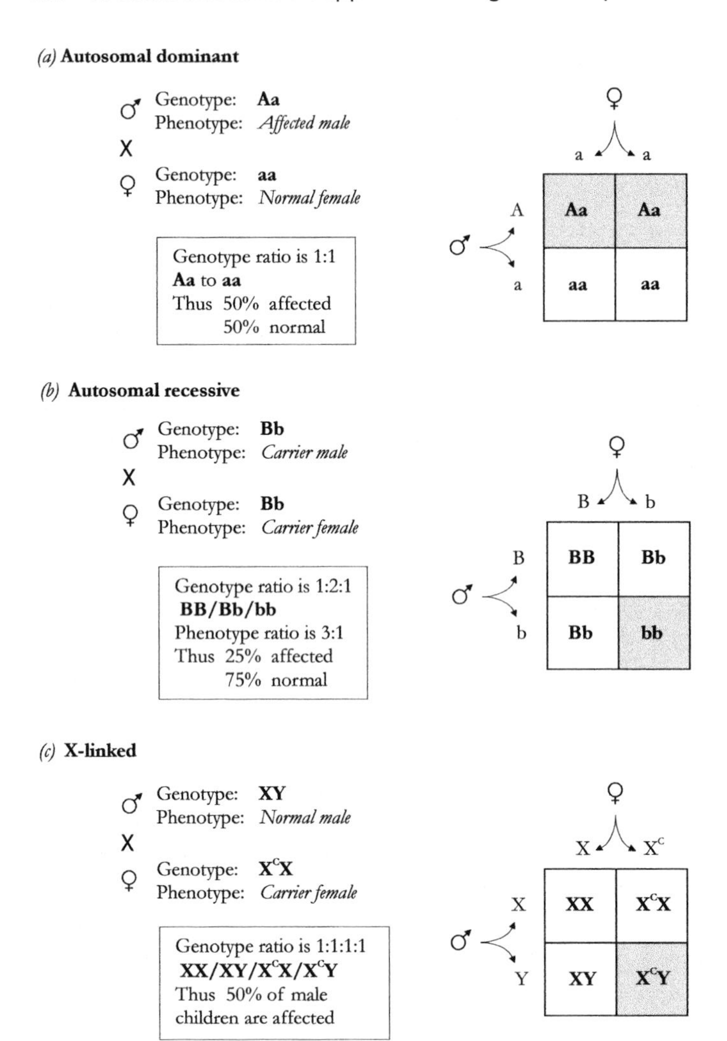

Fig. 11.1. Patterns of inheritance. In (*a*) an autosomal dominant disease allele is designated **A**, normal form **a**. Half of the gametes from an affected individual (in this case the male) will carry the disease allele. On mating (box diagram) the gametes can mix in the combinations shown. The result is that half the offspring will be heterozygous and therefore have the disease (shown by shaded boxes). In (*b*) an autosomal recessive pattern is shown. The disease-causing allele is designated **b**, the normal variant **B**. On a mating between two carriers heterozygous for the defective allele, there is a one in four chance of having an affected child. In (*c*) an X-linked pattern is shown for a disease allele designated **c**. In this case a recessive allele is shown. Half of the male children will be affected, as there is only one X chromosome and thus no dominant allele to mask the effect. No female children are affected. However, in the case of an X-linked dominant allele, females are also affected.

affect the range of phenotypes derived from what is actually a simple Mendelian pattern of inheritance. Further complications arise when **multiple alleles** are involved in determining traits, or when alleles demonstrate **incomplete dominance, co-dominance** or **partial dominance**. In many cases the route from genotype to phenotype also involves one or more environmental factors, when traits are said to be of a **multifactorial** nature.

Despite the complexities of transmission patterns and outcomes, there are many cases where the defect can be traced with reasonable certainty. As stated in Chapter 9, data for transmission of disease traits are collated in **Online Mendelian Inheritance in Man**, which now runs to over 12000 entries in various categories. The database is the electronic version of the text *Mendelian Inheritance in Man*, by Victor McKusick of Johns Hopkins University, who published the first edition in 1966. McKusick is rightly considered to be the father of medical genetics.

11.1.3 Genetically based disease conditions

Genetic problems may arise from either **chromosomal abnormalities** (**aberrations**) or **gene mutations**. An abnormal chromosome complement can involve whole chromosome sets (variation in the **ploidy** number, such as **triploid**, **tetraploid**, etc.) or individual chromosomes (**aneuploidy**). Any such variation usually has very serious consequences, often resulting in spontaneous abortion, as gene dosage is affected and many genes are involved. Multiple chromosome sets are rare in most animals, but are quite often found in plants. As gamete formation involves meiotic cell division in which homologous chromosomes separate during the reduction division, even-numbered multiple sets are most commonly found in polyploid plant species that remain stable.

Aneuploidy is a much more common form of chromosomal variation in humans, but is still relatively rare in terms of live-birth presentations. A missing chromosome gives rise to a **monosomic** condition, which is usually so severe that the foetus fails to develop fully. An additional chromosome gives a **trisomic** condition, which is more likely to persist to term. Monosomy and trisomy can affect both autosomes and sex chromosomes, with several recognised syndromes such as **Down syndrome** (trisomy-21). Most cases involving changes to chromosome number are caused by **non-disjunction** at meiosis during gamete formation. In addition to variation in chromosome number, structural changes can affect parts of chromosomes and can cause a range of conditions. Some examples of chromosomal aberrations in humans are shown in Table 11.1.

Table 11.1. *Some types of chromosomal aberration in humans*

Condition	Chromosome designation	Syndrome	Frequency per live births
Autosomal			
Trisomy-13	47, 13+	Patau syndrome	1:12 500–1:22 000
Trisomy-18	47, 18+	Edwards syndrome	1:6 000–1:10 000
Trisomy-21	47, 21+	Down syndrome	1:800
Sex chromosome variation			
Missing Y	45, X	Turner syndrome	1:3 000 female births
Additional X	47, XXX	Triplo-X	1:1 200 female births
Additional X	47, XXY	Klinefelter syndrome	1:500 male births
Additional Y	47, XYY	Jacobs syndrome	1:1 000 male births
Structural defects	**Cause**		
Deletion	Part of chromosome deleted, e.g. AB**CDE**FGH ABFGH		
Duplication	Part of chromosome duplicated, e.g. ABCDEFGH ABCD**BCD**EFGH		
Inversion	Part of chromosome inverted, e.g. ABCDEFGH ABC**FED**GH		
Translocation	Fragment moved to different chromosome, e.g. ABCDEFGH PQR**DEF**STUV		
Fragile-X syndrome	region of X-chromosome susceptible to breakage; known as Martin-Bell syndrome, presenting as 1:1 250 male births and 1:2 500 female births		

Note: Chromosome designation lists the total number of chromosomes, followed by the specific defect. Thus 47, 13+ indicates an additional chromosome 13, and 47, XXY a male with an additional X chromosome. The syndrome is usually named after the person who first described it; the possessive (e.g. *Down's* syndrome) is sometimes still used, but the modern convention is to use the non-possessive (e.g. *Down* syndrome).

Although chromosomal abnormalities are a very important type of genetic defect, it is in the characterisation of **gene mutations** that molecular genetics has had most impact. Many diseases have now been almost completely characterised, with their mode of transmission and action defined at both the chromosomal and molecular levels. Table 11.2 lists some of the more common forms of monogenic disorder that affect humans. We will consider some of these in more detail to outline how a disease can be characterised in terms of the effects of a mutated gene.

Cystic fibrosis (**CF**) is the most common genetically based disease found

Table 11.2. *Selected monogenic traits in humans*

Inheritance pattern/disease	Frequency per live births	Features of the disease condition
Autosomal recessive		
Cystic fibrosis	1:2 000–1:2 500 in Western Caucasians	Ion transport defects; lung infection and pancreatic dysfunction result
Tay–Sachs disease	1:3 000 in Ashkenazi Jews	Neurological degeneration, blindness and paralysis
Sickle-cell anaemia	1:50–1:100 in African populations where malaria is endemic	Sickle-cell disease affects red blood cells; heterozygous genotype confers a level of resistance to malaria
Phenylketonuria	1:2 000–1:5 000	Mental retardation due to accumulation of phenylalanine
α_1-Antitrypsin deficiency	1:5 000–1:10 000	Lung tissue damage and liver failure
Autosomal dominant		
Huntington disease	1:5 000–1:10 000	Late-onset motor defects, dementia
Familial hypercholesterolaemia	1:500	Premature susceptibility to heart disease
Breast cancer genes BRCA1 and 2	1:800 (1:100 in Ashkenazi Jews)	Susceptibility to early-onset breast and ovarian cancer
Familial retinoblastoma	1:14 000	Tumours of the retina
X-linked		
Duchenne muscular dystrophy	1:3 000–1:4 000	Muscle wastage, teenage onset
Haemophilia A/B	1:10 000	Defective blood clotting mechanism
Mitochondrial		
Leber hereditary optic neuropathy (LHON)	Mitochondrial defect, maternally inherited/late onset thus difficult to estimate	Optic nerve damage, may lead to blindness, but complex penetrance of the defective gene due to mitochondrial pattern of inheritance

in Western Caucasians, appearing with a frequency of around 1 in 2000–2500 live births. It is transmitted as an autosomal recessive characteristic, and therefore the birth of an affected child may be the first sign that there is a problem in the family. The carrier frequency for the CF defective allele is around 1 in 20–25 people. The disease presents with various symptoms, the most serious of which is the clogging of respiratory passageways with thick sticky mucus. This is too thick to be moved by the cilia that line the air passages, and the patient is likely to suffer persistent and repeated infections. Lung function is therefore compromised by CF, and even with improved treatments the life expectancy is only around 30 years. The pancreatic duct may also be affected by CF, resulting in **pancreatic exocrine deficiency**, which causes problems with digestion.

Cystic fibrosis can be traced in European folklore, from which the following puzzling statement comes: '*Woe to that child which when kissed on the forehead tastes salty. He is bewitched and soon must die*'. The condition was first described clinically in 1938, although characterisation of the disease at the molecular level was not achieved until the gene responsible was cloned in 1989. The defect responsible for CF affects a membrane protein involved in chloride ion transport, which results in epithelial cell sheets having insufficient surface hydration – hence the sticky mucus. There is also an increase in the salt content of sweat – hence the statement quoted above. The gene/protein responsible for CF is called the **cystic fibrosis transmembrane conductance regulator** (**CFTR**). So how was the gene cloned and characterised?

The hunt for the CF gene is a good example of how the technique of **positional cloning** can be used to find a gene for which the protein product is unknown, and for which there is little cytogenetic or linkage information available. Positional cloning, as the name suggests, involves identifying a gene by virtue of its position – essentially by deciphering the molecular connection between phenotype and genotype. In 1985, a linkage marker (called *met*) was found that localised the CF gene on the long arm of chromosome 7. The search for other markers uncovered two that showed no recombination with the CF locus – thus they were much closer to the CF gene. These markers were used as the start points for a trawl through some 280 kbp of DNA, looking for potential CF genes. This was done using the techniques of **chromosome walking** and **chromosome jumping** to search for contiguous DNA sequences from clone banks (this was before YAC and BAC vectors enabled large fragments to be cloned). The basis of chromosome walking and jumping is shown in Fig. 11.2. Using these methods, four candidate genes were identified from coding sequence information, and by tracing patterns of expression of each of these in CF patients, the search uncovered the 5′ end of a

(a) **Chromosome walking**

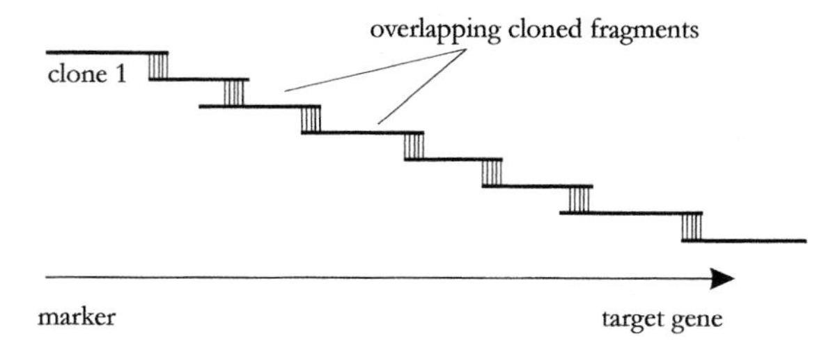

(b) **Chromosome jumping**

non-adjacent probes used to 'jump' along the chromosome

cloned region

jump

jump

unknown or unclonable region of the genome

walk

walk

Fig. 11.2. Chromosome walking and jumping. Chromosome walking (*a*) uses probes derived from the ends of overlapping clones to enable a 'walk' along the sequence. Thus a probe from clone 1 identifies the next clone, which then provides the probe for the next, and so on. In this way a long contiguous sequence can be assembled. In chromosome jumping (*b*), regions that are difficult to clone can be 'jumped'. The probes are prepared using a technique that enables fragments from distant sites to be isolated in a single clone by circularising a large fragment and isolating the region containing the original probe and the distant probe. This can then be used to isolate a clone containing sequences from the distant region. Often a combination of walks and jumps is needed to move from a marker (such as an RFLP) to a gene sequence. From Nicholl (2000), *Cell and Molecular Biology*, Advanced Higher Monograph Series, Learning and Teaching Scotland. Reproduced with permission.

large gene that was expressed in the appropriate tissues. This became known as the CFTR gene. A summary of the hunt for CFTR is shown in Fig. 11.3.

Having identified the CFTR gene, more detailed characterisation of its normal gene product, and the basis of the disease state, could begin. The gene is some 250 kbp in size, and encodes 27 exons which produce a protein of 1480 amino acids. The protein is similar to the ATP-binding cassette (ABC) family of membrane transporter proteins. When the gene was being characterised, it was noted that around 70% of CF cases appeared to have a similar

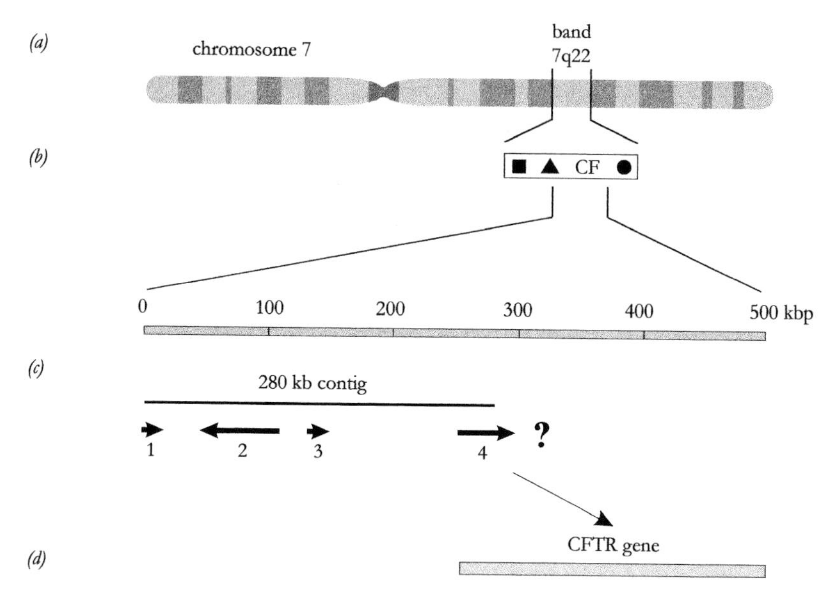

Fig. 11.3. The hunt for the cystic fibrosis gene. (*a*) Mapping studies placed the gene on the long arm of chromosome 7, at band position 7q22. (*b*) Markers associated with this region (square, triangle and circle) were mapped in relation to the CF gene. (*c*) A region of some 500 kbp was examined and a contiguous sequence (clone 'contig') of 280 kbp was identified. This region contained four candidate gene sequences or open reading frames (ORFs, labelled1–4). Further analysis of mRNA transcripts and DNA sequences eventually identified ORF 4 as the start of the 'CF gene', which was named the cystic fibrosis transmembrane conductance regulator (CFTR) gene.

defective region in the sequence – a 3 base-pair deletion in exon 10. This causes the amino acid **phenylalanine** to be deleted from the protein sequence. This mutation is called the *Δ*F508 mutation (*Δ* for deletion, F is single letter abbreviation for phenylalanine, and 508 is the position in the primary sequence of the protein). It affects the folding of the CFTR protein, which means that it cannot be processed and inserted into the membrane correctly after translation. Thus patients who carry two *Δ*F508 alleles do not produce any functional CFTR, with the associated disease phenotype arising as a consequence of this. The *Δ*F508 mutation is summarised in Fig. 11.4.

Molecular characterisation of a gene opens up the possibility of accurate diagnosis of disease alleles. Although screening for CF traditionally involved the 'sweat test', there is now a range of molecular techniques that can be used to confirm the presence of a defective CFTR allele, which can enable hetero-

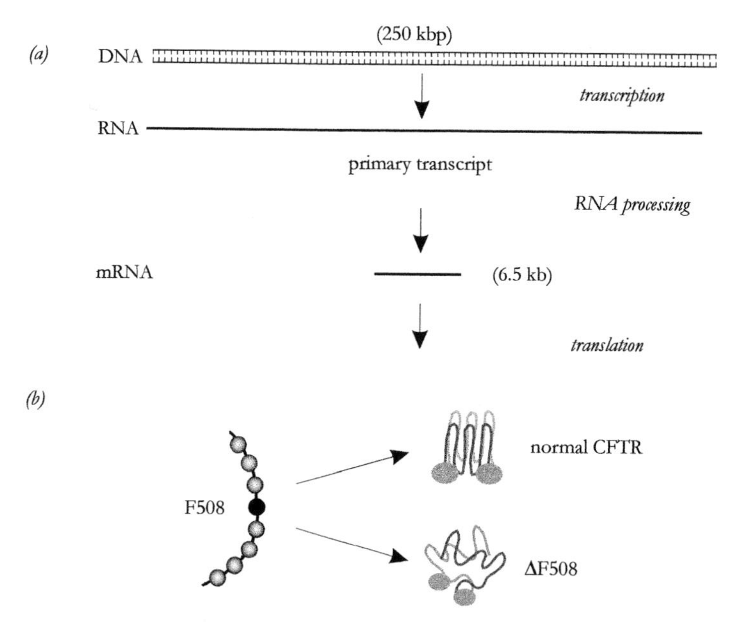

Fig. 11.4. The cystic fibrosis ΔF508 mutation. In (*a*) the gene is shown. Transcription produces the primary RNA transcript, that is converted into the functional 6.5 kb mRNA by removal of intervening sequences. On translation the transmembrane conductance regulator protein (CFTR) is produced. In (*b*) the normal and mutant proteins are shown. Normal CFTR has phenylalanine (*F*) at position 508. In the mutant ΔF508 protein this is deleted, causing the protein to fold incorrectly, which prevents it reaching the site of incorporation into the membrane. From Nicholl (2000), *Cell and Molecular Biology*, Advanced Higher Monograph Series, Learning and Teaching Scotland. Reproduced with permission.

zygous carriers to be identified with certainty. Two of these involve the use of PCR to amplify a fragment around the ΔF508 region to identify the 3-bp deletion, and the use of **allele-specific oligonucleotides** (**ASOs**) in hybridisation tests. The use of these techniques is shown in Fig. 11.5.

Although the ΔF508 mutation is the most common cause of CF, to date, around 1000 mutations have been identified in the CFTR gene. Many types of mutation have been characterised, including promoter mutations, frameshifts, amino acid replacements, defects in splicing, and deletions. With more sophisticated diagnosis, patients are being diagnosed with milder presentations of CF, which may not appear as early or be as severe as the ΔF508-based disease. Thus the CF story provides a good illustration of the scope of molecular biology in medical diagnosis, as it has enabled the common form of the disease to be characterised, and has also extended our knowledge of how *(a)* **Normal and mutant F508 sequences**

Ile Ile Phe Gly
Normal gene sequence 5' - GAA AAT ATC AT**C TT**T GGT GTT TCC - 3'

Mutant gene sequence 5' - GAA AAT ATC ATT GGT GTT TCC - 3'
Ile Ile Gly

(b) **PCR amplification of deleted region**

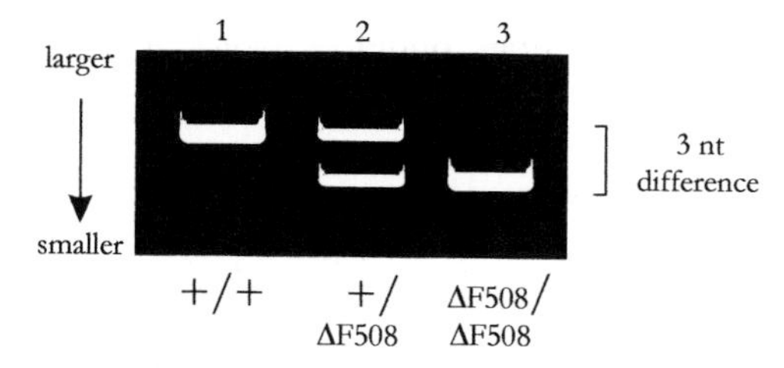

(c) **Using allele-specific oligonucleotide probes**

ASO 1 (normal) 3' - CTTTTATAGTAGAAACCACAAAGG - 5'
ASO 2 (mutant) 3' - CTTTTATAGTAACCACAAAGG - 5'

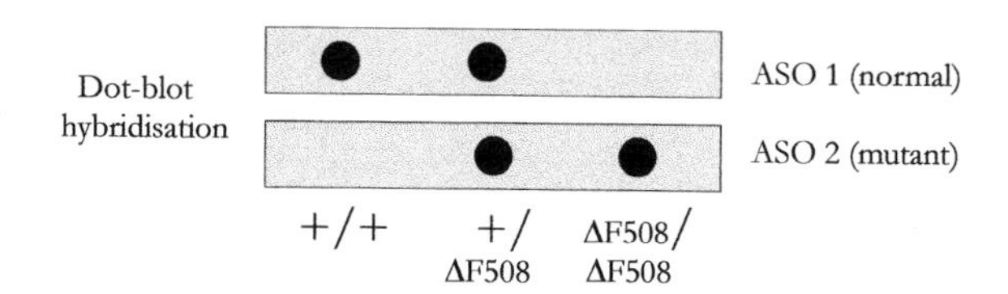

Fig. 11.5. Diagnosis of ΔF508 CF allele. In (*a*) the normal and mutant gene sequences around position 508 (Phe) are shown. The deleted 3 base-pairs are shaded in the normal sequence. This causes loss of phenylalanine. In (*b*) a PCR-based test is shown. A 100 base-pair region around the deletion is amplified using PCR, and the products run on a gel that will discriminate between the normal fragment and the mutant fragment, which will be three nucleotides smaller. Lanes 1, 2 and 3 show patterns obtained for homozygous normal (+/+), heterozygous carrier (+/ΔF508) and homozygous recessive CF patient (ΔF508/ΔF508). In (c) a similar pattern is seen with the use of allele-specific oligonucleotide probes (ASOs). The probe sequence is shown, derived from the gene sequences shown in (*a*). By amplifying DNA samples from patients using PCR and performing a dot-blot hybridisation with the radiolabelled ASOs, simple diagnosis is possible. In this example hybridisation with each probe separately enables the three genotypes to be determined by examining an autoradiograph.

highly polymorphic loci can influence the range of effects that may be caused by mutation.

In the area around Lake Maracaibo in Venezuela, there is a large family group of people who are descended from a woman who had migrated from Europe in the 1800s. Members of this group share a common ailment. They begin to exhibit peculiar involuntary movements, and also suffer from dementia and depression. Time of onset is usually around the age of 40–50. Their children, who were born when their parents were healthy, also develop the symptoms of this distressing condition, which is known as **Huntington disease** (**HD**; previously known as **Huntington's chorea**, which describes the *choreiform* movements of sufferers).

A clinical psychologist called Nancy Wexler has made a long-term study of thousands of HD sufferers from the Lake Maracaibo population, producing an extensive pedigree which confirms that HD follows an autosomal dominant pattern of inheritance, where the presence of a single defective allele is enough to trigger the disease state. Thus children of an affected parent have a 50% chance of inheriting the condition. As the disease presents with late onset (relative to child-bearing age), many people would wish to know if they carried the defective allele, so that informed choices could be made about having a family. The search for the gene responsible for HD involved tracing a restriction fragment length polymorphism (RFLP) that is closely linked to the HD locus. The RFLP, named G8, was identified in 1983. It segregates with the HD gene in 97% of cases. The HD gene itself was finally identified in 1993, located near the end of the short arm of chromosome 4. The defect involves a relatively unusual form of mutation called a **trinucleotide repeat**. The HD gene has multiple repeats of the sequence CAG, which codes for glutamine. In normal individuals, the gene carries up to 34 of these repeats. In HD alleles, more than 42 of the repeats indicates that the disease condition will appear. There is also a correlation between the number of repeats and the age of onset of the disease, which appears earlier in cases where larger numbers of repeats are present. As with cystic fibrosis, the availability of the gene sequence enables diagnostic tests to be developed for HD. Using PCR, the repeat region can be amplified and the products separated by gel electrophoresis to determine the number of repeats and thus the genetic fate of the individual with respect to HD.

Most X-linked gene disorders are recessive. However, their pattern of inheritance means that they are *effectively* dominant in males (XY) as there is no second allele present as would be the case for females (XX), or in an autosomal diploid situation. Thus X-linked diseases are often most serious in boys, as is the case for **muscular dystrophy** (**MD**). This is a muscle wasting disease that is progressive, usually from a teenage onset, and which causes premature death. The severe form of the disease is called **Duchenne muscular dystrophy** (**DMD**), although there is a milder form called **Becker muscular dystrophy** (**BMD**). Both these defects map to the same location on the X chromosome. The MD gene was isolated in 1987 using positional cloning techniques. It is extraordinarily large, covering 2.5 Mb of the X chromosome (that's 2500 kbp, or around 2% of the total!). The 79 exons in the MD gene produce a transcript of 14 kb, which encodes a protein of 3685 amino acids called **dystrophin**. Its function is to link the cytoskeleton of muscle cells to the sarcolemma (membrane).

The MD gene shows a much higher rate of mutation than is usual – some two orders of magnitude higher than other X-linked loci. This may simply be due to the extreme size of the gene, which therefore presents an 'easy target' for mutation. Most of the mutations characterised so far are deletions. Those that affect reading frame generally cause the severe Duchenne form of MD, whilst deletions that leave reading frame intact tend to be associated with the Becker form of the disease.

11.2 Treatment using rDNA technology – gene therapy

Once genetic defects have been identified and characterised, the possibility of treating the patient arises. If the defective gene can be replaced with a functional copy (sometimes called the **transgene**, as in transgenic) that is expressed correctly, the disease caused by the defect can be prevented. This approach is known as **gene therapy**, and is one of the most promising aspects of the use of gene technology in medicine. There are two possible approaches to gene therapy: (i) introduction of the transgene gene into the **somatic cells** of the affected tissue, or (ii) introduction into the reproductive (**germ line**) cells. These two approaches have markedly different ethical implications. Most scientists and clinicians consider somatic cell gene therapy an acceptable practice, no more morally troublesome than taking an aspirin. However, tinkering with the reproductive cells, with the probability of germ line transmission, is akin to altering the gene pool of the human species, which is regarded as unacceptable by most people. Thus genetic engineering of germ cells is an area that is likely to remain off limits at present.

There are several requirements for a gene therapy protocol to be effective. Firstly, the gene defect itself will have been characterised, and the gene cloned and available in a form suitable for use in a clinical programme. Secondly, there must be a system available for getting the gene into the correct site in the patient. Essentially these are vector systems that are functionally equivalent to vectors in a standard gene cloning protocol – their function is to carry the DNA sequence into the target cells. This also requires a mechanism for physical delivery to the target, which may involve inhalation, injection or other similar methods. Finally, if these requirements can be satisfied, the inserted gene must be expressed in the target cells if a non-functional gene is to be 'corrected'. Ideally, the faulty gene would be replaced by a functional copy. This is known as **gene replacement therapy**, and requires recombination between the defective gene and the inserted functional copy. Due to technical difficulties in achieving this reliably in target cells, the alternative is to use **gene addition therapy**. In addition therapy there is no absolute requirement for reciprocal exchange of the gene sequences, and the inserted gene functions alongside the defective gene. This approach is useful only if the gene defect is not dominant, in that a dominant allele will still produce the defective protein, which may overcome any effect of the transgene. Therapy for dominant conditions could be devised using **antisense mRNA**, in which a reversed copy of the gene is used to produce mRNA in the antisense configuration. This can bind to the mRNA from the defective allele and effectively prevent its translation. Antisense technology will be discussed in more detail in Chapter 12.

A further complication in gene therapy is the target cell or tissue system itself. In some situations it may be possible to remove cells from a patient and manipulate them outside the body. The altered cells are then replaced, with function restored. This approach is known as ***ex vivo* gene therapy**. It is mostly suitable for diseases that affect the blood system. It is not suitable for tissue-based diseases such as DMD of CF, in which the problem lies in dispersed and extensive tissue such as the lungs and pancreas (CF) or the skeletal muscles (DMD). It is difficult to see how these conditions could be treated by *ex vivo* therapy, and therefore the technique of treating these conditions at their locations is used. This is known as ***in vivo* gene therapy**. Features of these two types of gene therapy are illustrated in Fig. 11.6, with both approaches having been used with some success.

11.2.1 Getting transgenes into patients

Before looking at two examples of gene therapy procedures, it is worth reviewing the key methods available for getting the transgene into the cells of the patient. As we have seen, there are two aspects to this. The *biology* of the system must be established and evaluated, and then the *physical* method for

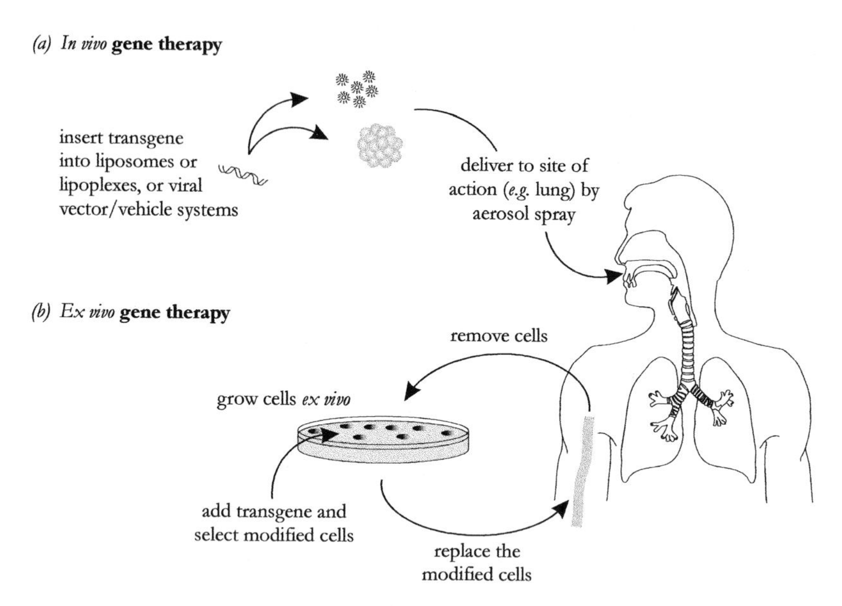

Fig. 11.6. *In vivo* and *ex vivo* routes for gene therapy. The *in vivo* approach is shown in (*a*). The gene is inserted into a vector or liposome/lipoplex, and introduced into target tissue of the patient. In this case the lung is the target, and an aerosol can be used to deliver the transgene. Such an approach can be used with cystic fibrosis therapy. In (*b*) the *ex vivo* route is shown. Cells (e.g. from blood or bone marrow) are removed from the patient and grown in culture medium. The transgene is therefore introduced into the cells outside the body. Modified cells can be selected and amplified (as in a typical gene-cloning protocol with mammalian cells) before they are injected back into the patient.

getting the gene to the site of action has to be considered. Deciding on the best method for addressing these two aspects of a therapeutic procedure is one important part of the strategy.

As with vectors for use in cloning procedures, viruses are an attractive option for delivering genes into human cells. We can use the term **vector** in its cloning context, as a piece of DNA into which the transgene is inserted. The viral particle itself is often called the **vehicle** for delivery of the transgene, although some authors describe the whole system simply as a vector system. Three main viral systems have been developed for gene therapy protocols, based on **retroviruses**, **adenoviruses** and **adeno-associated viruses**. The advantage of viral systems is that they provide a specific and efficient way of getting DNA into the target cells. However, care must be taken to ensure

Table 11.3. *Vector/vehicle systems for gene therapy*

System	Features
Viral-based	
Retroviruses	RNA genome, usually used with cDNA, requires proliferating cells for incorporation of the transgene into the nuclear material. Not specific for one cell type, and can activate cellular oncogenes
Adenoviruses	Double-stranded DNA genome, virus infects respiratory and gastrointestinal tract cells, thus effective in non- or slowly dividing cells. Generally provoke a strong immune response
Adeno-associated viruses	Replication-defective, thus requires helper virus. Some benefits over adenoviral systems, may show chromosome-specific integration of transgene
Non-viral	
Liposomes	System based on lipid micelles that encapsulate the DNA. Some problems with size, as micelles are generally small and may restrict the amount of DNA encapsulated. Inefficient compared to viral systems
Lipoplexes	Benefits over liposomes include increased efficiency due to charged groups present on the constituent lipids. Non-immunogenic, so benefits compared to viral systems
Naked DNA	Inefficient uptake but may be useful in certain cases

that viable virus particles are not generated during the therapy procedure, as this would potentially be detrimental to the patient.

In addition to viral-based systems, DNA can be delivered to target cells by non-viral methods. Naked DNA can be used directly, although this is not an efficient method. Alternatively, the DNA can be encapsulated in a lipid micelle called a **liposome**. Developments of this technique produced more complex structures that resemble viral particles, and these were given the name **lipoplexes** to distinguish them from liposomes. Some features of delivery systems are given in Table 11.3.

When a delivery system is available, the patient can be exposed to the virus in a number of ways. Delivery into the lungs by aerosol inhalation is one method appropriate to *in vivo* therapy for CF, as this is the main target tissue. Injection or infusion are other methods that may be useful, particularly if an *ex vivo* protocol has been used.

11.2.2 Gene therapy for adenosine deaminase deficiency

The first human gene therapy treatment was administered in September 1990 to a 4-year-old girl called Ashanti DaSilva, who received her own genetically altered white blood cells. Ashanti suffered from a recessive defect known as **adenosine deaminase (ADA) deficiency**, which causes the disease **severe combined immunodeficiency syndrome (SCIDS)**. Although a rare condition, this proved to be a suitable target for first steps in gene therapy in that the gene defect was known (the 32 kbp gene for ADA is located on chromosome 20), and an *ex vivo* strategy could be employed. Before gene therapy was available, patients could be treated by **enzyme replacement therapy**. A major development in this area was preparing the ADA enzyme with polyethylene glycol (PEG; the main component of antifreeze!) to stabilise delivery of the enzyme. The treatment is still important as an additional supplement to gene therapy, the response to which can be variable in different patients.

For the first ADA treatments, lymphocytes were removed from the patients and exposed to recombinant retroviral vectors to deliver the functional ADA gene into the cells. The lymphocytes were then replaced in the patients. Further developments came when bone marrow cells were used for the modification. The stem cells that produce T-lymphocytes are present in bone marrow, and thus altering these progenitor cells should improve the effect of the ADA transgene, particularly with respect to the duration of the effect. The problem is that T-lymphocyte stem cells are present as only a tiny fraction of the bone marrow cells, and thus efficient delivery of the transgene is difficult. Umbilical cord blood is a more plentiful source of target cells, and this method has been used to effect ADA gene therapy in newborn infants diagnosed with the defect. This approach is proving to be an effective therapy for this condition.

11.2.3 Gene therapy for cystic fibrosis

Cystic fibrosis is an obvious target for gene therapy, as it presents much more frequently than ADA deficiency, and is a major health problem for CF patients. Drug therapies can help to alleviate some of the symptoms of CF by digestive enzyme supplementation and the use of antibiotics to counter infection. However, as with ADA, enzyme replacement therapy is an attack on the *symptoms* of the disease, rather than on the *cause*. As outlined in Section 11.1.3, the defective gene/protein involved in CF has been defined and characterised. As CF is a recessive condition cause by a faulty protein, if a functional copy of the CFTR gene could be inserted into the appropriate tissue (chiefly the lung) then normal CFTR protein could be synthesised by the cells and thus restore the normal salt transport mechanism. Early indications that this could be achieved came from experiments that demonstrated that normal CFTR could be expressed in cell lines to restore defective CFTR function, thus opening up the real possibility of using this approach in patients.

Development of a suitable therapy for a disease such as CF usually involves developing an **animal model** for the disease, so that research can be carried out to mimic the therapy in a model system before it reaches clinical trials. In CF, the model was developed using transgenic mice that lack CFTR function. Adenovirus-based vector/vehicle systems were used, and these were shown to be effective. Thus the system seemed to be effective, and human trials could begin. In moving into a human clinical perspective, there are several things that need to be taken into account in addition to the science of the gene and its delivery system. For example, how can the efficacy of the technique be measured? As CF therapy involved cells deep in the lung, it is difficult to access these cells to investigate the expression of the normal CFTR transgene. Using nasal tissue can give some indications, but this is not completely reliable. Also, how effective must the transgene delivery/expression be in order to produce a clinically significant effect? Do all the affected cells in the lung have to be 'repaired', or will a certain percentage of them enable restoration of near-normal levels of ion transport?

Despite the problems associated with devising, applying and monitoring gene therapy for CF, there have been many clinical trials to date, with some success achieved. Both viral-based and liposome/lipoplex delivery systems have been used, although the application of CF gene therapy by a widely available, robust and effective method is still relatively distant. However, progress is being made at a great rate, and many scientists believe that an effective therapy is within reach.

11.3 DNA profiling

Given the size of the human genome, and our knowledge of genome structure, it is relatively easy to calculate that each person's genome is unique, the only exceptions being **monozygotic twins** (twins derived from a single fertilised ovum). This provides the opportunity to use the genome as a unique identifier, if suitable techniques are available to generate robust and unambiguous results. The original technique was called **DNA fingerprinting**, but with improved technology the range of tests that can be carried out has increased, and today the more general term **DNA profiling** is preferred. The technique has found many applications in both criminal cases and in disputes over whether people are related or not (paternity disputes and immigration cases are the most common). The basis of all the techniques is that a sample of DNA from a suspect (or person in a paternity or immigration dispute) can be matched with that of the reference sample (from the victim of a crime, or a relative in a civil case). In scene-of-crime investigations, the technique can be limited by the small amount of DNA available is forensic samples. Modern techniques use the polymerase chain reaction to amplify and detect minute samples of DNA from bloodstains, body fluids, skin fragments or hair roots.

11.3.1 The history of 'genetic fingerprinting'

The original DNA fingerprinting technique was devised by Alec Jeffreys in 1985, who realised that the work he was doing on sequences within the myoglobin gene could have wider implications. The method is based on the fact that there are highly variable regions of the genome that are specific to each individual. These are **minisatellite** regions, which have a variable number of short repeated-sequence elements known as **variable number tandem repeats** (**VNTRs**, see Chapter 9, and Fig. 9.8). Within the VNTR there are core sequence motifs that can be identified in other polymorphic VNTR loci, and also sequences that are restricted to the particular VNTR. The arrangement of the VNTR sequences, and the choice of a suitable probe sequence, are the key elements that enable a unique 'genetic fingerprint' to be produced.

The first requirement is to isolate DNA and prepare restriction fragments for electrophoresis. As shown in Fig. 9.8, if an enzyme is used that does *not* cut the core sequence, but cuts frequently outside it, then the VNTR is effectively isolated. For human DNA the enzyme *Hin*fI is often used. By using a probe that hybridises to the core sequence, and carrying out the hybridisation under low stringency, polymorphic loci that bind the probe can be identified. The probe in this case is known as a **multi-locus probe**, as it binds to multiple sites. This generates a pattern of bands that is unique – the 'genetic fingerprint'. If probes with sequences that are specific for a particular VNTR are used (**single-locus probes**), a more restricted fingerprint is produced, as there will be two alleles of the sequence in each individual, one maternally derived and one paternally derived. An overview of the basis of the technique is shown in Fig. 11.7.

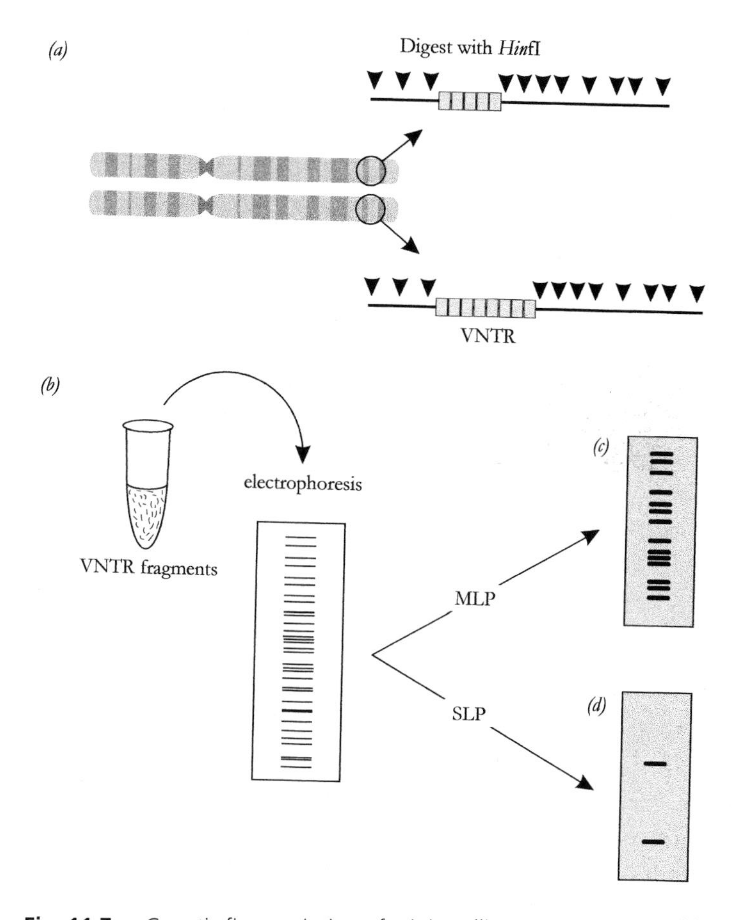

Fig. 11.7. Genetic fingerprinting of minisatellite DNA sequences. (*a*) shows a chromosome pair, with one minisatellite (VNTR) locus highlighted. In this case the locus is heterozygous for VNTR length. Cutting with *Hin*fI effectively isolates the VNTR. In (*b*) the VNTR fragments produced (from many loci) are separated by electrophoresis and blotted. Challenging with a multi-locus probe (MLP) produces the 'bar code' pattern shown in (*c*). If a single-locus probe (SLP) is used, the two alleles of the specific VNTR are identified as shown in (*d*).

In forensic analysis, the original DNA profiling technique has now been largely replaced by a PCR-based method that amplifies parts of the DNA known as **short tandem repeats** (**STRs**, also known as **microsatellites**). These are repeats of 2, 3, 4 or 5 base-pairs. A major advantage over minisatellite (VNTR) repeats is that STR repeats are found throughout the genome, thus better coverage is achieved than with minisatellites. The PCR reaction overcomes any problems associated with the tiny amounts of sample that are often found at the crime scene. The reaction is set up to amplify the loci involved – usually three or four are sufficient if the loci are selected carefully to optimise the information generated. By using fluorescent labels and automated DNA detection equipment (similar to the genome sequencing equipment shown in Fig. 9.6) a DNA profile can be generated quickly and accurately.

11.3.2 DNA profiling and the law

The use of DNA profiling is now accepted as an important way of generating evidence in legal cases. In addition to the science itself, which may involve multi-locus or single-locus probes, or (more usually) STR amplification, there are several factors that must be considered if the evidence is to be sound. The techniques must be reliable, and must be accessible to trained technical staff, who must be aware of the potential problems with the use of DNA profiling. To ensure that results from DNA profile analysis are admissible as evidence in legal cases, rigorous quality control measures must be in place. These include accurate recording of the samples as they arrive at the laboratory, and careful cross-checking of the procedures to make sure that the test is carried out properly and that the samples do not get mixed up. If PCR amplification is used as part of the procedure, great care must be taken to ensure that no trace of DNA contamination is present. A smear of the operator's sweat can often be enough to ruin a test, so strict operating procedures must be observed, and laboratories inspected and authorised to conduct the tests. This is essential if public confidence in the technique is to be maintained.

An important consideration in legal cases is the likelihood of matching DNA profiles being generated by chance from two different individuals. This is obviously critical in cases where legal decisions are made on the strength of DNA fingerprint evidence, and perhaps custodial sentences passed. Although there is no dispute about the fact that we all have unique genomes, DNA profiling of course can only examine a small part of the genome. Thus the odds of a chance match need to be calculated. The more bands present in a DNA profile, the less likely a non-related match will be found. The odds against a chance match for varying numbers of bands in a DNA profile are shown in Table 11.4. It is generally accepted that an approved DNA profiling agency, working under agreed conditions with standardised protocols, will generate results that are valid and reliable.

The use of multi-locus DNA profiling in a forensic case is shown in Fig. 11.8. In this example, blood from the victim is the reference sample. Samples

Table 11.4. *The odds against chance matches in a DNA fingerprint*

Number of bands in fingerprint	Odds against a chance match
4	250 : 1
6	4000 : 1
8	65 000 : 1
10	1 million : 1
12	17 million : 1
14	268 million : 1
16	4300 million : 1
18	68 000 million : 1
20	1 million million : 1

Note: The more bands present in a DNA fingerprint, the less likely it is that any match is due to chance. However, allele frequencies for different genes may have to be taken into account. Allele frequencies can vary in different populations, and again this may be important in a legal situation. Generally, problems can be avoided by taking all known factors into account, and assessing the risk of a chance match by taking the highest estimate. (Data courtesy of Cellmark Diagnostics, reproduced with permission.)

from seven suspects were obtained and treated along with the sample from the victim. By matching the band patterns it is clear that suspect 5 is the guilty party.

Single-locus probes will bind to just one complementary sequence in the haploid genome. Thus, two bands will be visible in the resulting autoradiogram: one from the paternal chromosome and one from the maternal chromosome. This gives a simple profile that is often sufficient to demonstrate an unambiguous match between the suspect and the reference. The result of a paternity test using a single-locus probe is shown in Fig. 11.9. Sometimes, two or more probes can be used to increase the number of bands in the profile. Single-locus probes are more sensitive than multi-locus probes, and can detect much smaller amounts of DNA. Usually both single-locus and multi-locus probes are used in any given case, and the results combined.

11.3.3 Mysteries of the past revealed by genetic detectives

Another interesting development and application of rDNA technology has been in examining the past – a sort of 'genetic history' trail. Although the

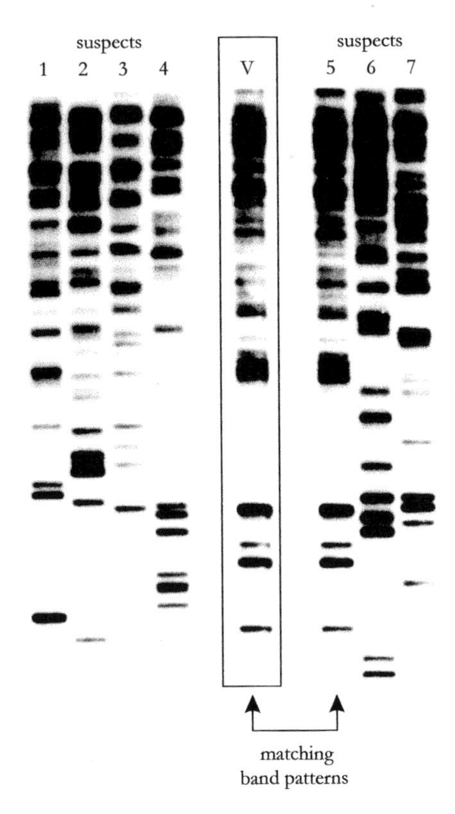

Fig. 11.8. A DNA profile prepared using a multi-locus probe. Samples of the suspect's DNA isolated from the victim (V; boxed) and seven candidate suspects (1–7) were cut with a restriction enzyme and separated on an agarose gel. The fragments were blotted onto a filter and challenged with a radioactive probe. The probe hybridises to the target sequences, producing a profile pattern when exposed to X-ray film. The band patterns from the victim's sample and suspect 5 match. (Courtesy of Cellmark Diagnostics. Reproduced with permission.)

book (and subsequent film) *Jurassic Park* perhaps were a little too fanciful, it is surprising how some of the ideas portrayed in fiction have been used in the real world.

Identifying individuals using DNA profiling can go much further back in time than the immediate past, where perhaps a recently deceased individual is identified by DNA analysis. In the 1990s, DNA analysis enabled the identification of notable people such as **Josef Mengele** (of Auschwitz notoriety) and **Czar Nicholas II** of Russia, thus solving long-running debates about their deaths. DNA has also been extracted from an Egyptian mummy (2400

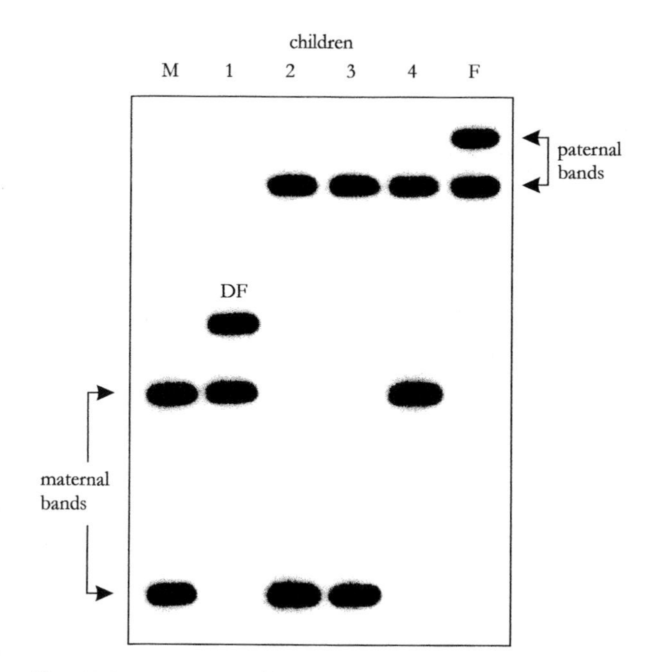

Fig. 11.9. A DNA profile prepared using a single-locus probe for paternity testing. Samples of DNA from the mother (M), four children (1–4) and the father (F) were prepared as in Fig. 11.8. A single-locus probe was used in this analysis. The band patterns therefore show two maternal bands and two paternal bands, one band from each homologous chromosome on which the target sequence is located. In the case of child 1, the paternal band is different from either of the two bands in lane F, indicating a different father (band labelled DF). This child was in fact born to the mother during a previous marriage. Courtesy of Cellmark Diagnostics. Reproduced with permission.

years old), and a human bone that is 5500 years old! Thus it is possible to use rDNA technology to study ancient DNA, recovered from museum specimens or newly discovered archaeological material. Topics such as the migration of ancient populations, the degree of relatedness between different groups of animals, and the evolution of species can be addressed if there is access to DNA samples that are not too degraded. This area of work is sometimes called **molecular paleontology**. With DNA having been extracted from fossils as old as 65 million years, the 'genetics of the past' looks like providing evolutionary biologists, taxonomists and paleontologists with much useful information in the future.

Use of human remains in the identification of disease-causing organisms can also be a fruitful area of research. For example, there was some debate as to the source of tuberculosis in the Americas – did it exist before the early explorers reached the New World, or was it a 'gift' from them? By analysing DNA from the lung tissue of a Peruvian mummy, researchers found DNA that corresponded to the tubercule bacillus *Mycobacterium tuberculosis*, thus proving that the disease was, in fact, endemic in the Americas prior to the arrival of the European settlers.

In addition to its use in forensic and legal procedures, and in tracing genetic history, DNA profiling is also a very powerful research tool that can be applied in many different contexts. Techniques such as RAPD analysis and genetic (fingerprinting) are being used with many other organisms, such as cats, dogs, birds and plants. Application of the technique in an ecological context enables problems that were previously studied by classical ecological methods to be investigated at the molecular level. This use of **molecular ecology** is likely to have a major impact on the study of organisms in their natural environments, and (like molecular paleontology) is a good example of the coming together of branches of science that were traditionally treated as separate disciplines.

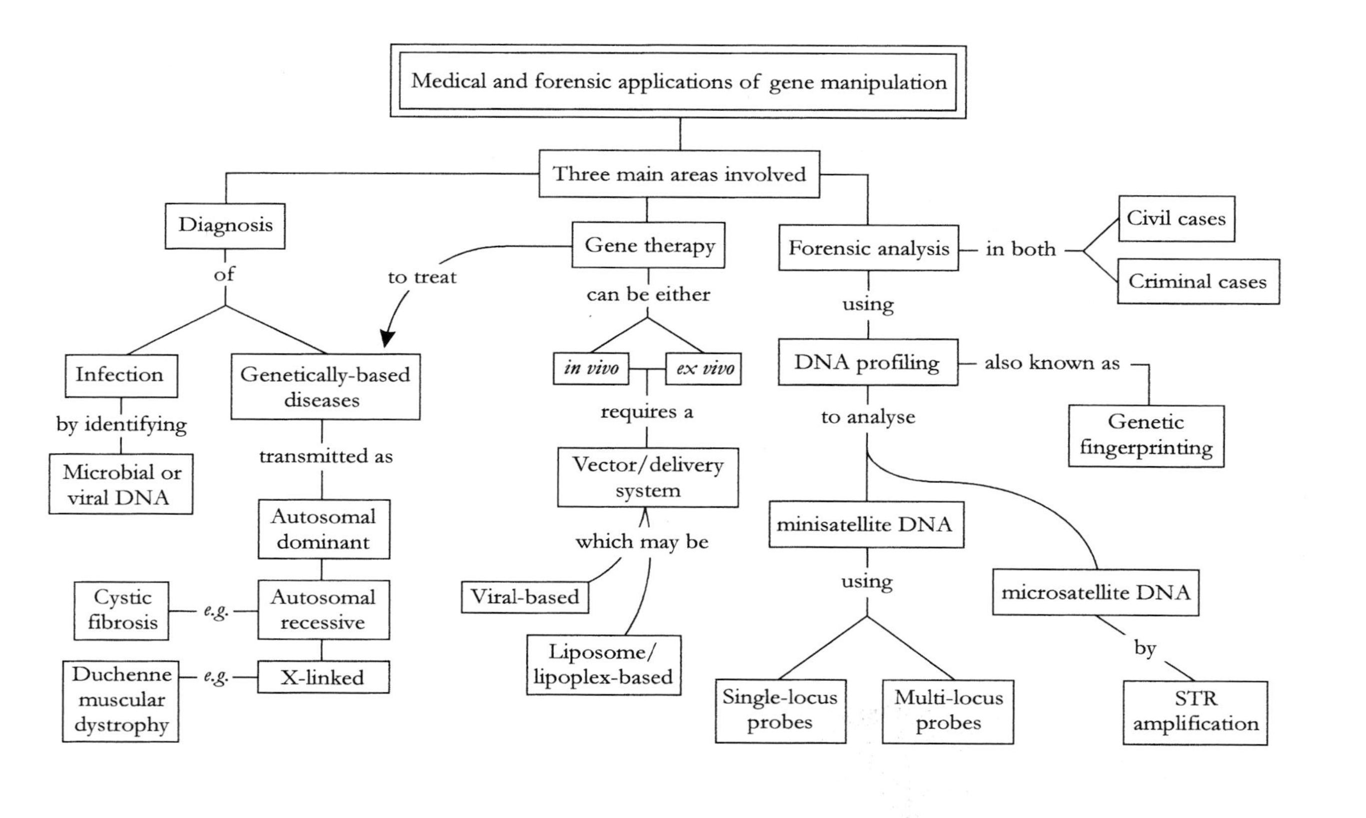

Concept map 11

12

Transgenic plants and animals

The production of a **transgenic** organism involves altering the genome so that a permanent change is effected. This is different from somatic cell gene therapy, in which the effects of the transgene are restricted to the individual who receives the treatment. In fact, the whole point of generating a transgenic organism is to alter the germ line, so that the genetic change is inherited in a stable pattern following reproduction. This is one area of genetic engineering that has caused great public concern, and there are many complex issues surrounding the development and use of transgenic organisms. In addition, the scientific and technical problems associated with genetic engineering in higher organisms are often difficult to overcome. This is partly due to the size and complexity of the genome, and partly due to the fact that the development of plants and animals is an extremely complex process that is still not yet fully understood at the molecular level. Despite these difficulties, methods for the generation of transgenic plants and animals are now well established, and the use of transgenic organisms has already had a major impact in a range of different disciplines. In this chapter we will consider the development and use of transgenic plants and animals.

12.1 Transgenic plants

All life on earth is dependent on the photosynthetic fixation of carbon dioxide by plants. We sometimes lose sight of this fact, as most people are removed from the actual process of generating our food, and the supermarket shelves have all sorts of exotic processed foods and pre-prepared meals that somehow swamp the vegetable section. Despite this, the generation of transgenic plants, particularly in the context of **genetically modified foods**, has produced an enormous public reaction to an extent that no-one could have predicted. We will return to this debate in Chapter 14. In this section, we will look at the science of transgenic plant production.

12.1.1 Why transgenic plants?

For thousands of years, humans have manipulated the genetic characteristics of plants by selective breeding. This approach has been extremely successful, and will continue to play a major part in agriculture. However, classical plant breeding programmes rely on being able to carry out genetic crosses between individual plants. Such plants must be sexually compatible (which usually means that they have to be closely related), and thus it has not been possible to combine genetic traits from widely differing species. The advent of genetic engineering has removed this constraint, and has given the agricultural scientist a very powerful way of incorporating defined genetic changes into plants. Such changes are often aimed at improving the productivity and 'efficiency' of crop plants, both of which are important to help feed and clothe the increasing world population.

There are many diverse areas of plant genetics, biochemistry, physiology and pathology involved in the genetic manipulation of plants. Some of the prime targets for the improvement of crop plants are listed in Table 12.1. In many of these, success has already been achieved to some extent. However, many people are concerned about the possible ecological effects of the release of genetically modified organisms (**GMOs**) into the environment, and there is much debate about this aspect. The truth of the matter is that we simply do not know what the consequences might be – a very small alteration to the balance of an ecosystem, caused by a more vigorous or disease-resistant plant, might have a considerable knock-on effect over a long timescale.

As in other areas, successful genetic manipulation of plants requires: (i) methods for introducing genes into plants, and (ii) a detailed knowledge of the molecular genetics of the system that is being manipulated. In many cases the latter is the limiting factor, particularly where the characteristic under study involves many genes (a **polygenic** trait). However, despite the problems, plant genetic manipulation is already having a considerable impact on agriculture.

Table 12.1. *Possible targets for crop plant improvement*

Target	Benefit(s)
Disease, Herbicide, Insect, Virus resistance	Improve productivity of crops and reduce their loss due to biological agents
Cold, Drought, Salt tolerance	Permit growth of crops in areas that are physically unsuitable at present
Reduction of photorespiration	Increase efficiency of energy conversion
Nitrogen fixation	Confer ability to fix atmospheric nitrogen to a wider range of species
Nutritional value	Improve nutritional value of storage proteins by protein engineering
Storage properties	Extend shelf-life of fruits and vegetables
Consumer appeal	Make fruits and vegetables more appealing with respect to colour, shade, size, etc.

12.1.2 Ti plasmids as vectors for plant cells

Introducing cloned DNA into plant cells is now routine practice in many laboratories worldwide. A number of methods can be used to achieve this, including physical methods such as microinjection or biolistic DNA delivery. Alternatively, a biological method can be used in which the cloned DNA is incorporated into the plant by a vector. Although plant viruses such as **calumoviruses** or **geminiviruses** may be attractive candidates for use as vectors, there are several problems with these systems. Currently, the most widely used plant cell vectors are based on the Ti plasmid of *Agrobacterium tumefaciens*, which is a soil bacterium that is responsible for **crown gall disease**. The bacterium infects the plant through a wound in the stem, and a tumour of cancerous tissue develops at the crown of the plant.

The agent responsible for the formation of the crown gall tumour is not the bacterium itself, but a plasmid known as the **Ti plasmid** (Ti stands for **T**umour **i**nducing). Ti plasmids are large, with a size range in the region of 140 kb to 235 kb. In addition to the genes responsible for tumour formation, the Ti plasmids carry genes for virulence functions and for the synthesis and util-

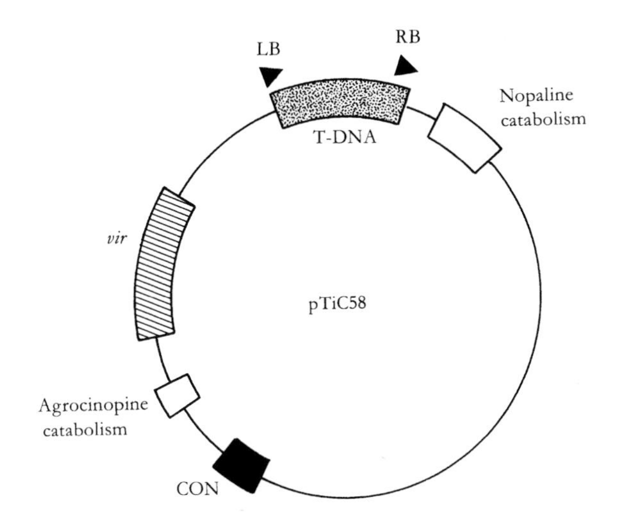

Fig. 12.1. Map of the nopaline plasmid pTiC58. Regions indicated are the T-DNA (shaded), which is bordered by left and right repeat sequences (LB and RB); the genes for nopaline and agrocinopine catabolism, and the genes specifying virulence (*vir*). The CON region is responsible for conjugative transfer. From Old and Primrose (1989), *Principles of Gene Manipulation*, Blackwell. Reproduced with permission.

isation of unusual amino acid derivatives known as **opines**. Two main types of opine are commonly found, these being **octopine** and **nopaline**, and Ti plasmids can be characterised on this basis. A map of a nopaline Ti plasmid is shown in Fig. 12.1.

The region of the Ti plasmids responsible for tumour formation is known as the **T-DNA**. This is some 15–30 kb in size, and also carries the gene for octopine or nopaline synthesis. On infection, the T-DNA becomes integrated into the plant cell genome, and is therefore a possible avenue for the introduction of foreign DNA into the plant genome. Integration can occur at many different sites in the plant genome, the choice being apparently random. Nopaline T-DNA is a single segment, whereas octopine DNA is arranged as two regions known as the left and right segments. The left segment is similar in structure to nopaline T-DNA, and the right is not necessary for tumour formation. The structure of nopaline T-DNA is shown in Fig. 12.2. Genes for tumour morphology are designated *tms* ('shooty' tumours), *tmr* ('rooty' tumours) and *tml* ('large' tumours). The gene for nopaline synthase is designated *nos* (in octopine T-DNA this is *ocs*, encoding octopine synthase). The *nos* and *ocs* genes are eukaryotic in character, and their promoters have been used widely in the construction of vectors that express cloned sequences.

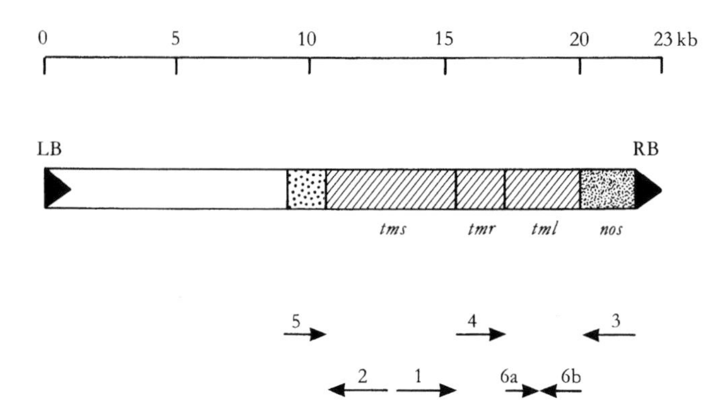

Fig. 12.2. Map of the nopaline T-DNA region. The left and right borders are indicated by LB and RB. Genes for nopaline synthase (*nos*) and tumour morphology (*tms*, *tmr* and *tml*) are shown. The transcript map is shown below the T-DNA map. Transcripts 1 and 2 (*tms*) are involved in auxin production, transcript 4 (*tmr*) in cytokinin production. These specify either shooty or rooty tumours. Transcript 3 encodes nopaline synthase, and transcripts 5 and 6 encode products that appear to supress differentiation. From Old and Primrose (1989), *Principles of Gene Manipulation*, Blackwell. Reproduced with permission.

Ti plasmids are too large to be used directly as vectors, so smaller vectors have been constructed that are suitable for manipulation in vitro. These vectors do not contain all the genes required for Ti-mediated gene transfer, and thus have to be used in conjunction with other plasmids to enable the cloned DNA to become integrated into the plant cell genome. Often a **tripartite** or **triparental** cross is required, where the recombinant is present in one *E. coli* strain, and a conjugation-proficient plasmid in another. A Ti plasmid derivative is present in *A. tumefaciens*. When the three strains are mixed, the conjugation-proficient 'helper' plasmid transfers to the strain carrying the recombinant plasmid, which is then mobilised and transferred to the *Agrobacterium*. Recombination then permits integration of the cloned DNA into the Ti plasmid, which can transfer this DNA to the plant genome on infection.

12.1.3 Making transgenic plants

In the development of transgenic plant methodology, two approaches using Ti-based plasmids were devised: (i) cointegration and (ii) the binary vector system. In the cointegration method, a plasmid based on pBR322 is used to

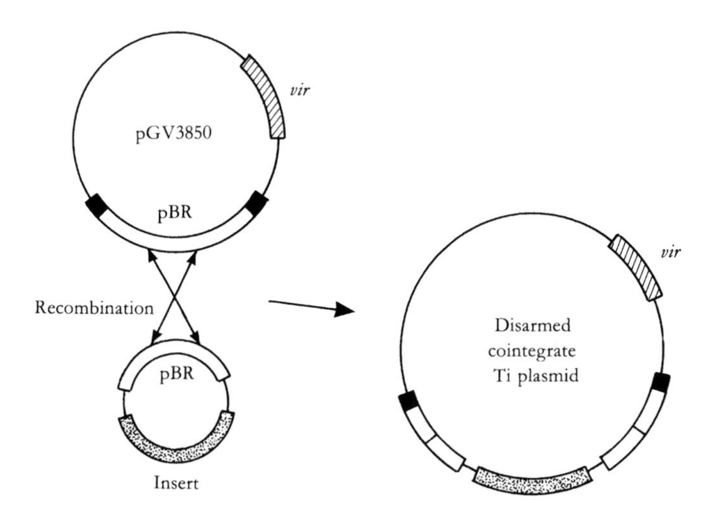

Fig. 12.3. Formation of a cointegrate Ti plasmid. Plasmid pGV3850 carries the *vir* genes, but has had some of the T-DNA region replaced with pBR322 sequences (pBR). The left and right borders of the T-DNA are present (filled regions). An insert (shaded) cloned into a pBR322-based plasmid can be inserted into pGV3850 by homologous recombination between the pBR regions, producing a disarmed cointegrate vector.

clone the gene of interest (Fig. 12.3). This plasmid is then integrated into a Ti-based vector such as pGV3850. This carries the *vir* region (which specifies virulence), and has the left and right borders of T-DNA, which are important for integration of the T-DNA region. However, most of the T-DNA has been replaced by a pBR322 sequence, which permits incorporation of the recombinant plasmid by homologous recombination. This generates a large plasmid that can facilitate integration of the cloned DNA sequence. Removal of the T-DNA has another important consequence, as cells infected with such constructs do not produce tumours and are subsequently much easier to regenerate into plants by tissue culture techniques. Ti-based plasmids lacking tumourigenic functions are known as **disarmed** vectors.

The binary vector system uses separate plasmids to supply the disarmed T-DNA (**mini-Ti** plasmids) and the virulence functions. The mini-Ti plasmid is transferred to a strain of *A. tumefaciens* (which contains a compatible plasmid with the *vir* genes) by a triparental cross. Genes cloned into mini-Ti plasmids are incorporated into the plant cell genome by *trans* complementation, where the *vir* functions are supplied by the second plasmid (Fig. 12.4).

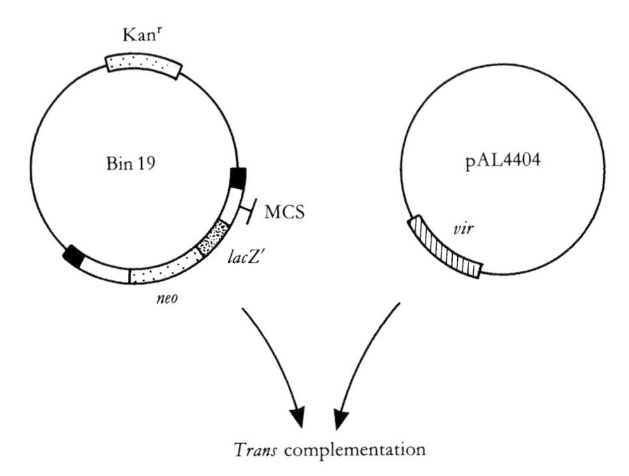

Fig. 12.4. Binary vector system based on Bin 19. The Bin 19 plasmid carries the gene sequence for the α-peptide of β-galactosidase (*lacZ'*), downstream from a polylinker (MCS) into which DNA can be cloned. The polylinker/*lacZ'*/*neo* region is flanked by the T-DNA border sequences (filled regions). In addition genes for neomycin phosphotransferase (*neo*) and kanamycin resistance (Kan^r) can be used as selectable markers. The plasmid is used in conjunction with pAL4404, which carries the *vir* genes but has no T-DNA. The two plasmids complement each other in *trans* to enable transfer of the cloned DNA into the plant genome. From Old and Primrose (1989), *Principles of Gene Manipulation*, Blackwell. Reproduced with permission.

When a suitable strain of *A. tumefaciens* has been generated, containing a disarmed recombinant Ti-derived plasmid, infection of plant tissue can be carried out. This is often done using leaf discs, from which plants can be regenerated easily, and many genes have been transferred into plants by this method. The method is summarised in Fig. 12.5. The one disadvantage of the Ti system is that it does not normally infect monocotyledonous (monocot) plants such as cereals and grasses. As many of the prime target crops are monocots, this has hampered the development with these varieties. However, other methods (such as direct introduction or the use of biolistics) can be used to deliver recombinant DNA to the cells of monocots, thus avoiding the problem.

12.1.4 Putting the technology to work

Transgenic plant technology has been used for a number of years, with varying degrees of success. One of the major problems we have already men-

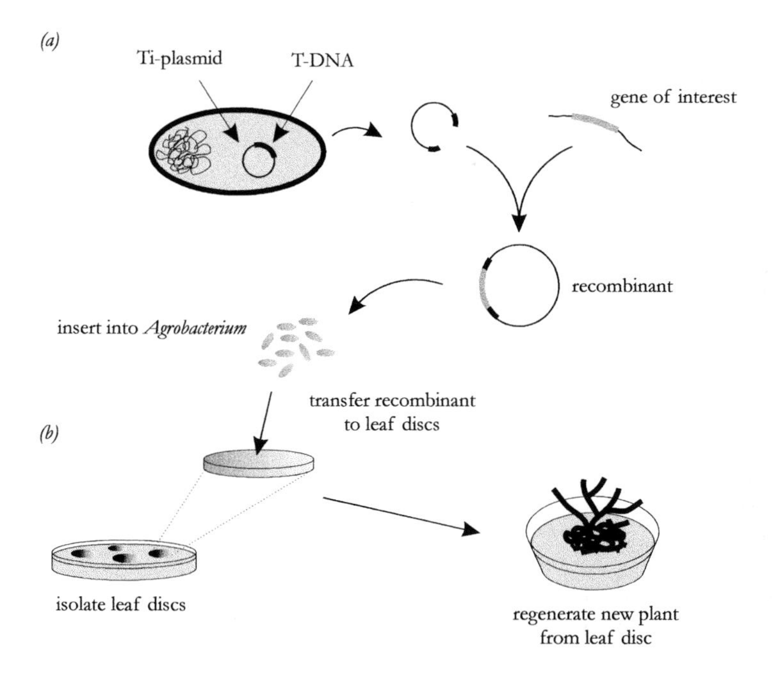

Fig. 12.5. Regeneration of transgenic plants from leaf discs. In (*a*) the target gene is cloned into a vector based on the Ti plasmid. The recombinant plasmid is used to transform *Agrobacterium tumefaciens* (often a triparental cross is used as described in the text). The bacterium is then used to infect plant cells that have been grown as disc explants of leaf tissue, as shown in (*b*). The transgenic plant is regenerated from the leaf disc by propagation on an appropriate tissue culture medium.

tioned has been the public acceptance of transgenic plants. In many cases the resistance from the public has been more of a problem than the actual science, and towards the end of 1999 the backlash against so-called 'Frankenfoods' had reached the point where companies involved were being affected, either directly by way of action against field trials, etc., or indirectly in that consumers were not buying the transgenic products. We will examine this area in more depth in Chapter 14.

One of the first recombinant DNA experiments to be performed with plants did not, in fact, produce transgenic plants at all, but involved the use of genetically modified bacteria. In nature, ice often forms at low temperatures by associating with proteins on the surface of so-called **ice-forming bacteria**, which are associated with many plant species. One of the most common ice-forming bacterial species is *Pseudomonas syringae*. In the late 1970s and early 1980s researchers removed the gene that is responsible for synthesising the ice-forming protein, producing what became known as **ice-minus bacteria**. Plans to spray the ice-minus strain onto plants in field trials were ready by 1982, but approval for this first **deliberate release experiment** was delayed as the issue was debated. Finally, approval was granted in 1987, and the field trial took place. Some success was achieved as the engineered bacteria reduced frost damage in the test treatments.

The bacterium *Bacillus thuringiensis* has been used to produce transgenic plants known as **Bt plants**. The bacterium produces toxic crystals that kill caterpillar pests when they ingest the toxin. The bacterium itself has been used as an insecticide, sprayed directly onto crops. However, the gene for toxin production has been isolated and inserted into plants such as corn, cotton, soybean and potato, with the first Bt crops planted in 1996. By 2000 over half of the soybean crop in the USA was planted with Bt-engineered plants, although there have been some problems with pests developing resistance to the Bt toxin. Monsanto have been active in developing Bt crops, as shown in Table 12.2.

One concern that has been highlighted by the planting of Bt corn is the potential risk to non-target species. In 1999 a report in *Nature* suggested that larvae of the Monarch butterfly, widely distributed in North America, could be harmed by exposure to Bt corn pollen, even though the regulatory process involved in approving the Bt corn has examined this possibility and found no significant risk. Risk is associated with both toxicity and exposure, and subsequent research has demonstrated that exposure levels are likely to be too low to pose a serious threat to the butterfly. However, the debate continues. This example illustrates the need for more extensive research in this area, if genetically modified crops are to gain full public acceptance.

Herbicide resistance is one area where a lot of effort has been directed. The theory is simple – if plants can be made herbicide-resistant, then weeds can be treated with a broad-spectrum herbicide without the crop plant being affected. One of the most common herbicides is **glyphosate**, which is available commercially as **Roundup** and **Tumbleweed**. Glyphosate acts by inhibiting an amino acid biosynthetic enzyme called **5-enolpyruvylshikimate-3-phosphate synthase** (**EPSP synthase** or **EPSPS**). Resistant plants have been produced by either increasing the synthesis of EPSPS by incorporating extra copies of the gene, or by using a bacterial EPSPS gene that is slightly different from the plant version, and produces a protein that is resistant to the effects of glyphosate. Monsanto have produced various crop plants, such as soya, that are called **Roundup-ready**, in that they are resistant to the herbicide. Such plants are now used widely in the USA and some other countries, as shown in Table 12.3.

Tomatoes are usually picked green, so that they are able to withstand

Table 12.2. *Established Bt crops*

Trade name	Countries planting	Area planted in 1999 (ha)
YieldGard corn	USA	6500000
	Canada	121000
	Argentina	17000
BollGard cotton	USA	1500000
	Argentina	10000
	Australia	75000
	Mexico	17000
NewLeaf potatoes	Field trials	N/A

Note: Areas are hectares for 1999 plantings. YieldGard, BollGard and NewLeaf are trade names of Monsanto. Data are from Monsanto website [www.monsanto.com], reproduced with permission.

Table 12.3. *Established Roundup Ready crops*

Trade name	Countries planting	Area planted in 1999 (ha)
Roundup Ready soybeans	USA	15000000
	Canada	142000
	Argentina	6000000
	Mexico	520 (limited distribution)
	Romania	14000
Roundup Ready corn	USA	930000
	Canada	8000
	Bulgaria	12000
Roundup Ready canola	USA	N/A
(rapeseed)	Canada	N/A

Note: Areas are hectares for 1999 plantings. Roundup Ready is a trade name of Monsanto. Data are from Monsanto website [www.monsanto.com], reproduced with permission.

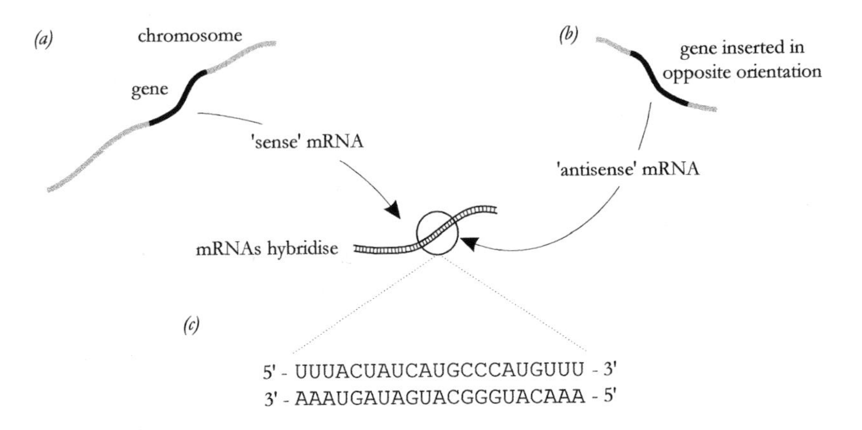

Fig. 12.6. Antisense technology. The target gene is shown in (*a*). A copy of the gene is introduced into a separate site on the genome, but in the opposite orientation, as shown in (*b*). On transcription of the antisense gene, an antisense mRNA is produced. This binds to the normal mRNA, preventing translation. Part of the sequence is shown in (*c*) to illustrate.

shipping and transportation without bruising. They are then ripened artificially by using ethylene gas, as ethylene is a key trigger for the ripening process. In trying to delay natural ripening, two approaches have been used. One is to target the production of ethylene itself, thus delaying the onset of the normal ripening mechanism. A second approach illustrates how a novel idea, utilising advanced gene technology to achieve an elegant solution to a defined problem, can still fail due to other considerations – this is the story of the **Flavr Savr** (*sic*) tomato.

The biotechnology company Calgene developed the Flavr Savr tomato using what became known as **antisense technology**. In this approach, a gene sequence is inserted in the opposite orientation, so that on transcription an mRNA that is complementary to the normal mRNA is produced. This antisense mRNA will therefore bind to the normal mRNA in the cell, inhibiting its translation and effectively shutting off expression of the gene. The principle of the method is shown in Fig. 12.6. In the Flavr Savr, the gene for the enzyme **polygalacturonase** (**PG**) is the target. This enzyme digests pectin in the cell wall, and leads to fruit softening and the onset of rotting. The elegant theory is that inhibition of PG production should slow the decay process, thus the fruit should be easier to handle and transport after picking. It can also be left on the vine to mature longer than is usually the case, thus improving flavour. After much development, the Flavr Savr became the first genetically modified food to be approved for use in the USA, in 1994. The level of PG was reduced to something like 1% of the normal levels, and the product seemed to be set for commercial success. However, various problems with the characteristics affecting growth and picking of the crop led to the failure of the Flavr Savr in commercial terms. Calgene is now part of Monsanto, having been stretched too far by development of the Flavr Savr. Despite the failure of the Flavr Savr, the company has continued to produce innovations in biotechnology, including rapeseed oil (known as canola oil in the USA) with a high concentration of lauric acid, which is beneficial from a health perspective.

Attempts to improve the nutritional quality of crops is not restricted to the commercial or healthfood sectors. For many millions of people around the world, access to basic nutrition is a matter of survival rather than choice. Rice is the staple food of some 3 billion people, and about 10% of these suffer from health problems associated with vitamin A deficiency. It is estimated that around 1 million children die prematurely due to this deficiency, with a further 350000 going blind. Thus rice has been one of the most intensively studied crop species with respect to improving quality of life for around half of the world's population. This led to the development of 'Miracle Rice', which was a product of the green revolution of the 1960s. However, widespread planting throughout South East Asia led to a rice monoculture, with increased susceptibility to disease and pests, and the increased dependence on pesticides that this brings. Thus, as with genetically modified crops, there can be problems in adopting new variants of established crop species.

In 1999 Ingo Potrykus, working in Zurich, succeeded in producing rice with **β-carotene** in the grain endosperm, where it is not normally found. As β-carotene is a precursor of vitamin A, increasing the amount available by engineering rice in this way should help to alleviate some of the problems of vitamin A deficiency. This is obviously a positive development. However, corporate interests in patent rights to the technologies involved, and other non-scientific problems, had to be sorted out before agreement was reached that developing countries could access the technology freely. Development work is continuing, but once again it clear that it is difficult to strike a balance between commercial factors and the potential benefits of transgenic plant technology. This is a particularly sensitive topic when those who would benefit most may not have the means to afford the technology.

A further twist in the corporate vs. common good debate can be seen in the so-called **gene protection technology**. This is where companies design their systems so that their use can be controlled, by some sort of manipulation that is essentially separate from the actual transgenic technology that they are designed to deliver. This has caused great concern among public and pressure groups, and a vigorous debate has ensued between such groups and the companies who are developing the technology. Even the terms used to describe the various approaches reflect the strongly held views of proponents and opponents. Thus one system is called a **technology protection system** by the corporate sector, and **terminator technology** by others. Another type of system is called **genetic use restriction technology** (**GURT**) or **genetic trait control technology**; this is also known more widely as **traitor technology**. So what do these emotive terms mean, and is there any need for concern?

Terminator technology is where plants are engineered to produce seeds that are essentially sterile, or do not germinate properly. Thus the growers are prevented from gathering seeds from one year to plant the next season, and are effectively tied to the seed company, as they have to buy new seeds each year. Companies reasonably claim that they have to obtain returns on the considerable investment required to develop a transgenic crop plant. Others, equally reasonably, insist that this constraint places poor farmers in developing countries at a considerable disadvantage, as they are unable to save seed from one year's crop to develop the next season's planting. Approximately half of the world's farmers are classed as 'poor' and cannot afford to buy new seed each season. They produce around 20% of world food output, and feed some 1.4 billion people. Thus there is a major ethical issue surrounding the prevention of seed-collecting from year to year, if terminator technology were to be applied widely. There is also great concern that any sterility-generating technology could transfer to other variants, species or genera, thus having a devastating effect on third-world farming communities.

Traitor technology or GURT involves the use of a 'switch' (often controlled by a chemical additive) to permit or restrict a particular engineered trait. This is perhaps a little less contentious than terminator technology, as the aim is to regulate a particular modification, rather than to prevent viable seed production. However, there are still many who are very concerned about the potential uses of this technology, which could again tie growers to a particular company if the 'switch' requires technology that only that company can supply.

Some of the major agricultural biotechnology companies have stated publicly that they will not develop terminator technology, which is seen as a partial success for the negative reaction of the public, farmers and pressure groups. There is, however, still a lot of uncertainty in this area, with some groups claiming that traitor technology is being developed further, and that corporate mergers can negate promises made previously by one of the partners in the merger. The whole field of transgenic plant technology is therefore in some degree of turmoil, with many conflicting interests, views and personalities involved. The debates are set to continue for a long time to come.

12.2 Transgenic animals

The generation of transgenic animals is one of the most complex aspects of genetic engineering, in terms both of technical difficulty and in the ethical problems that arise. Many people, who accept that the genetic manipulation of bacterial, fungal and plant species is beneficial, find difficulty in extending this acceptance when animals (particularly mammals) are involved. The need for sympathetic and objective discussion of this topic by the scientific community, the media and the general public is likely to present one of the great challenges in scientific ethics over the next few years.

12.2.1 Why transgenic animals?

Genetic engineering has already had an enormous impact on the study of gene structure and expression in animal cells, and this is one area that will continue to develop. Cancer research is one obvious example, and current investigation into the molecular genetics of the disease requires extensive use of gene manipulation technology. In the field of protein production in biotechnology (discussed in Chapter 10), the synthesis of many mammalian-derived recombinant proteins is often best carried out using cultured mammalian cells, as these are sometimes the only hosts which will ensure the correct expression of such genes.

Cell-based applications such as those outlined above are an important part of genetic engineering in animals. However, the term **transgenic** is usually reserved for whole organisms, and the generation of a transgenic animal is much more complex than working with cultured cells. Many of the problems have been overcome using a variety of animals, with early work involving amphibians, fish, mice, pigs and sheep.

Transgenics can be used for a variety of purposes, covering both basic research and biotechnological applications. The study of embryological development has been extended by the ability to introduce genes into eggs or early embryos, and there is scope for the manipulation of farm animals by the incorporation of desirable traits via transgenesis. The use of whole organisms for the production of recombinant protein is a further possibility, and this has already been achieved in some species. The term **pharm animal** or **pharming** (from **pharm**aceutical) is sometimes used when talking about the production of high-value therapeutic proteins using transgenic animal technology.

When considered on a global scale, the potential for exploitation of transgenic animals would appear to be almost unlimited. Achieving that potential is likely to be a long and difficult process in many cases, but the rewards are such that a considerable amount of money and effort has already been invested in this area.

12.2.2 Producing transgenic animals

There are several possible routes for the introduction of genes into embryos, each with its own advantages and disadvantages. Some of the methods are: (i) direct transfection or retroviral infection of embryonic stem cells followed by introduction of these cells into an embryo at the blastocyst stage of development, (ii) retroviral infection of early embryos, (iii) direct microinjection of DNA into oocytes, zygotes or early embryo cells, (iv) sperm-mediated transfer, (v) transfer into unfertilised ova, and (vi) physical techniques such as biolistics or electrofusion. In addition to these methods, the technique of **nuclear transfer** (used in organismal cloning, discussed in Chapter 13) is sometimes associated with a transgenesis protocol.

Early success was achieved by injecting DNA into one of the **pronuclei** of a fertilised egg, just prior to the fusion of the pronuclei (which produces the diploid zygote). This approach led to the production of the celebrated 'supermouse' in the early 1980s, which represents one of the milestones of genetic engineering. The experiments which led to the 'supermouse' involved placing a copy of the rat growth hormone (GH) gene under the control of the mouse metallothionine (mMT) gene promoter. To create the 'supermouse', a linear fragment of the recombinant plasmid carrying the fused gene sequences (MGH) was injected into the male pronuclei of fertilised eggs (linear fragments appear to integrate into the genome more readily than circular sequences). The resulting fertilised eggs were implanted into the uteri of foster mothers, and some of the mice resulting from this expressed the growth hormone gene. Such mice grew some two to three times faster than control mice, and were up to twice the size of the controls. Pronuclear microinjection is summarised in Fig. 12.7.

In generating a transgenic animal, it is desirable that all the cells in the organism receive the transgene. Presence of the transgene in the germ cells of the organism will enable the gene to be passed on to succeeding generations, and this is essential if the organism is to be useful in the long term. Thus introduction of genes has to be carried out at a very early stage of development, ideally at the single-cell zygote stage. If this cannot be achieved, there

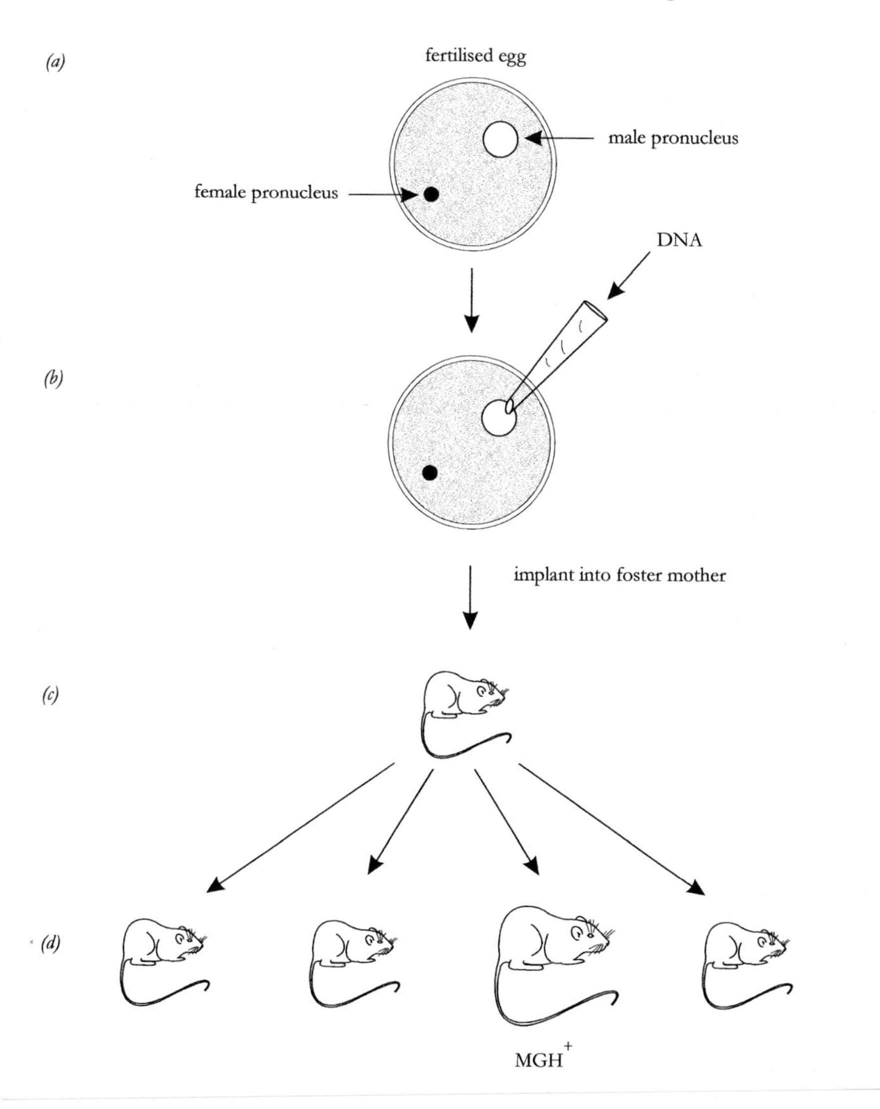

Fig. 12.7. Production of 'supermouse'. (*a*) Fertilised eggs were removed from a female and (*b*) the DNA carrying the rat growth hormone gene/mouse metallothionine promoter construct (MGH) was injected into the male pronucleus. (*c*) The eggs were then implanted into a foster mother. (*d*) Some of the pups expressed the MGH construct (MGH^+), and were larger than the normal pups.

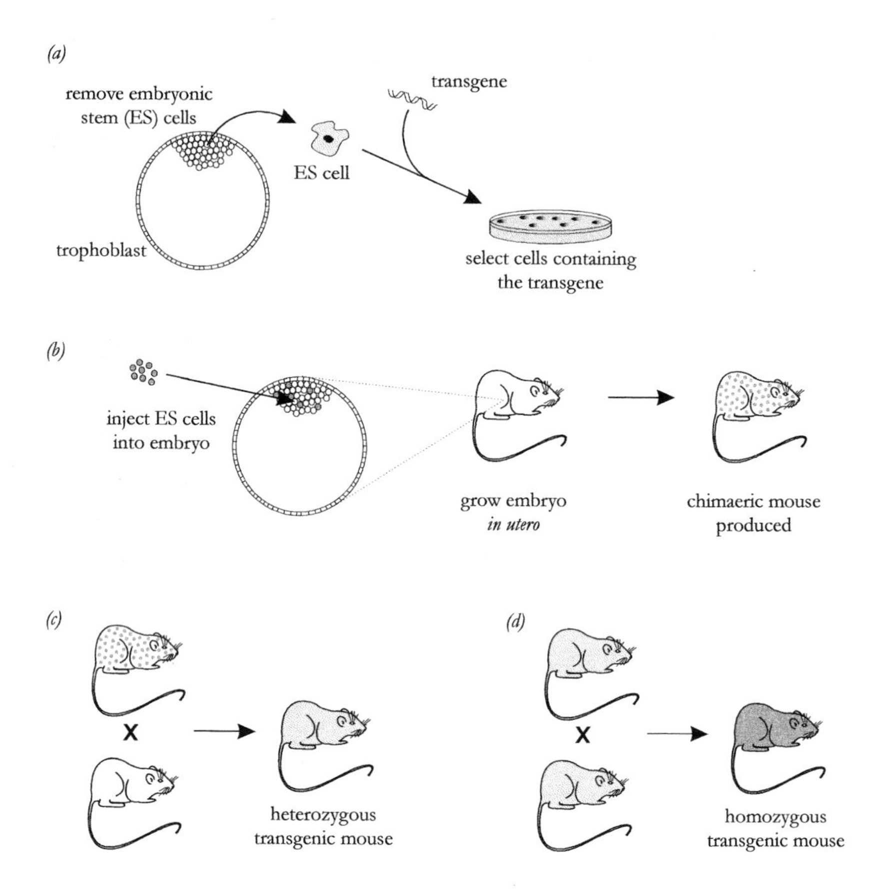

Fig. 12.8. Production of transgenic mice using embryonic stem cell technology. (*a*) Embryonic stem cells (ES cells) are removed from an early embryo and cultured. The target transgene is inserted into the ES cells, which are grown on selective media. (*b*) The ES cells containing the transgene are injected into the ES cells of another embryo, where they are incorporated into the cell mass. The embryo is implanted into a pregnant mother, and a chimaeric transgenic mouse is produced. By crossing the chimaera with a normal mouse as shown in (*c*), some heterozygous transgenics will be produced. If they are then self-crossed, homozygous transgenics will be produced in about 25% of the offspring, as shown in (*d*).

is the possibility that a **mosaic** embryo will develop, in which only some of the cells carry the transgene. Another example of this type of variation is where the embryo is generated from two distinct individuals, as is the case when embryonic stem cells are used. This results in a **chimaeric** organism. In practice this is not necessarily a problem, as the organism can be crossed to produce offspring that are homozygous for the transgene in all cells. A chimaeric organism that contains the transgene in its germ-line cells will pass the gene on to its offspring, which will therefore be heterozygous for the transgene (assuming they have come from a mating with a homozygous non-transgenic). A further cross with a sibling will result in around 25% of the offspring being homozygous for the transgene. This procedure is outlined for the mouse in Fig. 12.8.

12.2.3 Applications of transgenic animal technology

Introduction of growth hormone genes into animal species has been carried out, notably in pigs, but in many cases there are undesirable side effects. Pigs with the bovine growth hormone gene show greater feed efficiency and have lower levels of subcutaneous fat than normal pigs. However, problems such as enlarged heart, high incidence of stomach ulcers, dermatitis, kidney disease and arthritis have demonstrated that the production of healthy transgenic farm animals is a difficult undertaking. Although progress is being made, it is clear that much more work is required before genetic engineering has a major impact on animal husbandry.

The study of development is one area of transgenic research that is currently yielding much useful information. By implanting genes into embryos, features of development such as tissue-specific gene expression can be investigated. The cloning of genes from the fruit fly *Drosophila melanogaster*, coupled with the isolation and characterisation of transposable elements (P elements) that can be used as vectors, has enabled the production of stable transgenic *Drosophila* lines. Thus the fruit fly, which has been a major contributor to the field of classical genetic analysis, is now being studied at the molecular level by employing the full range of gene manipulation techniques.

In mammals, the mouse is proving to be one of the most useful model systems for investigating embryological development, and the expression of many transgenes has been studied in this organism. One such application is shown in Fig. 12.9, which demonstrates the use of the *lacZ* gene as a means of detecting tissue-specific gene expression. In this example the *lacZ* gene was placed under the control of the weak thymidine kinase (TK) promoter from herpes simplex virus (HSV), generating an HSV-TK–*lacZ* construct. This was used to probe for active chromosomal domains in the developing embryos, with one of the transgenic lines showing the brain-specific expression seen in Fig. 12.9.

Although the use of transgenic organisms is providing many insights into

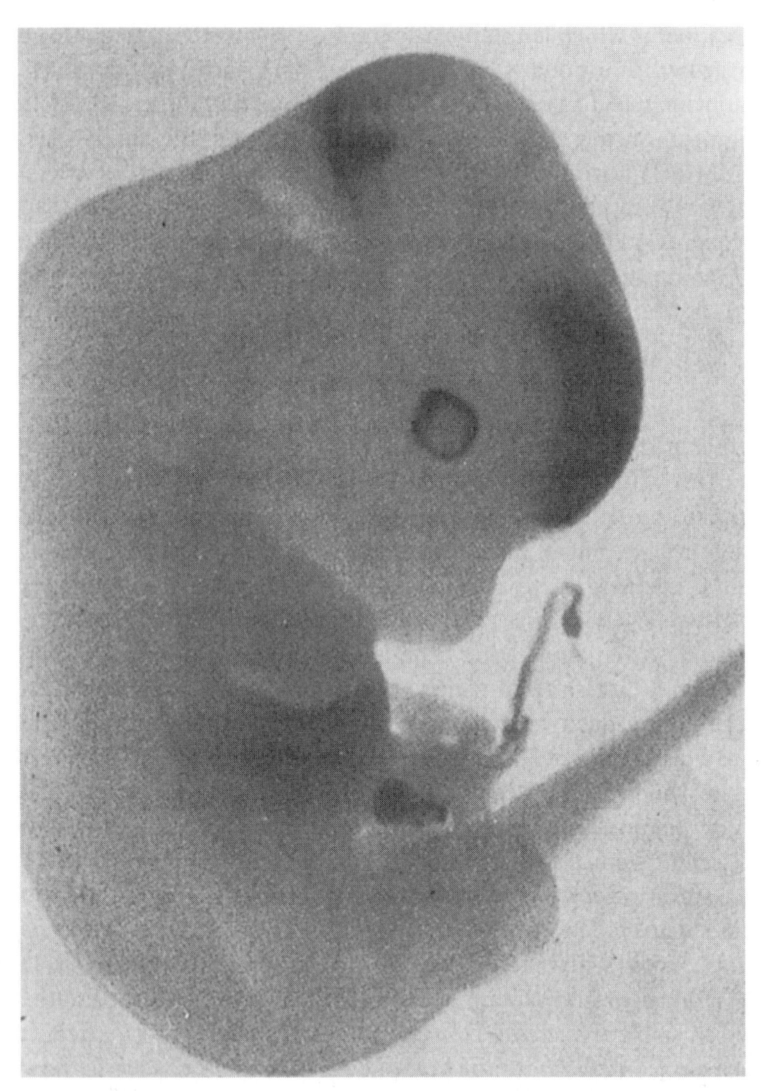

Fig. 12.9. Expression of a transgene in the mouse embryo. The β-galactosidase (*lacZ*) coding region was placed under the control of a thymidine kinase promoter from herpes simplex virus to produce the HSV-TK–*lacZ* gene construct. This was injected into male pronuclei and transgenic mice produced. The example shows a 13-day foetus from a transgenic strain that expresses the transgene in brain tissue during gestation. Detection is by the blue colouration produced by the action of β-galactosidase on X-gal. Thus the dark areas in the fore- and hind-brain are regions where the *lacZ* gene has been expressed. Photograph courtesy of Dr S. Hettle. From Allen *et al.* (1988), *Nature (London)* **333**, 852–855. Copyright (1988) Macmillan Magazines Limited. Reproduced with permission.

developmental processes, inserted genes may not always be expressed in exactly the same way as would be the case in normal embryos. Thus a good deal of caution is often required when interpreting results. Despite this potential problem, transgenesis is proving to be a powerful tool for the developmental biologist.

Mice have also been used widely as animal models for disease states. One celebrated example is the **oncomouse**, generated by Philip Leder and his colleagues at Harvard University. Mice were produced in which the ***c-myc* oncogene** and sections of the **mouse mammary tumour (MMT) virus** gave rise to breast cancer. The oncomouse has a place in history as the first complex animal to be granted a patent in the USA. Other transgenic mice with disease characteristics include the **prostate mouse** (prostate cancer), mice with **severe combined immunodeficiency syndrome (SCIDS)**, and mice that show symptoms of **Alzheimer disease**.

Many new variants of transgenic mice have been produced, and have become an essential part of research into many aspects of human disease. Increased knowledge of molecular genetics, and the continued development of the techniques of transgenic animal production, have enabled mice to be generated in which specific genes can be either activated or inactivated. Where a gene is inactivated or replaced with a mutated version, a **knockout** mouse is produced. If an additional gene function is established, this is sometimes called a **knockin** mouse. The use of knockout mice in cystic fibrosis (CF) research is one example of the technology being used in both basic research and in developing gene therapy procedures. Mice have been engineered to express mutant CF alleles, including the prevalent ΔF508 mutation that is responsible for most serious CF presentations. Having a mouse model enables researchers to carry out experiments that would not be possible in humans, although (as with developmental studies) results may not be exactly the same as would be the case in a human subject.

In 1995 a database was established to collate details of knockout mice. This is known as **MKMD** (**Mouse Knockout & Mutation Database**) and can be found at URL [**http://research.bmn.com/mkmd/**]. At the time of writing, some 5000 entries were listed, representing over 2000 unique genes, with more being added daily. This demonstrates that the mouse is proving to be one of the mainstays of modern transgenic research.

Early examples of protein production in transgenic animals include expression of human **tissue plasminogen activator** (**tPA**) in transgenic mice, and of human **blood coagulation factor IX** (**FIX**) in transgenic sheep. In both cases the transgene protein product was secreted into the milk of lactating organisms by virtue of being placed under the control of a milk-protein gene promoter. In the mouse example the construct consisted of the regulatory sequences of the **whey acid protein (WAP)** gene, giving a WAP–tPA construct. Control sequences from the **ß-lactoglobulin (BLG)** gene were used to generate a BLG–FIX construct for expression in the transgenic sheep. Other examples of transgenic animals acting as bioreactors include pigs that express human haemoglobin, and cows that produce human lactoferrin.

Producing a therapeutic protein in milk provides an ideal way of ensuring a reliable supply from lactating animals, and downstream processing to obtain purified protein is relatively straightforward. This approach was used by scientists from the Roslin Institute near Edinburgh, working in conjunction with the biotechnology company PPL Therapeutics. In 1991 PPL's first transgenic sheep (called Tracy) was born. Yields of human proteins of around 40 g l^{-1} were produced from milk, demonstrating the great potential for this technology. PPL continues to develop a range of products, such as **α_1-antitrypsin** (for treating cystic fibrosis), **fibrinogen** (for use in medical procedures) and **human factor IX** (for haemophilia B).

Using animals as bioreactors offers an alternative to the fermentation of bacteria or yeast that contain the target gene. The technology is now becoming well established, with many biotechnology companies involved. As we have already seen, this area is sometimes called **pharming**, with the transgenic animals referred to as **pharm animals**. Given the problems in achieving correct expression and processing of some mammalian proteins in non-mammalian hosts, this is proving to be an important development of transgenic animal technology.

Transgenic animals also offer the potential to develop organs for **xenotransplantation**. The pig is the target species for this application, as the organs are of similar size to human organs. In developing this technology the key target is to alter the cell surface recognition properties of the donor organs, so that the transplant is not rejected by the human immune system. In addition to whole organs from mature animals, there is the possibility of growing tissue replacements as an additional part of a transgenic animal. Both these aspects of xenotransplantation are currently being investigated, with a lot of scientific and ethical problems still to be solved. Given the shortfall in organ donors, and the consequent loss of life or quality of life that results, many people feel that the objections to xenotransplantation must be discussed openly, and overcome, to enable the technology to be implemented when fully developed.

Our final word on transgenic animals brings the technology closer to humans. In January 2001 the birth of the first transgenic non-human primate, a rhesus monkey, was announced. This was developed by scientists in Portland, Oregon. He was named ANDi, which stands for inserted DNA (written backwards!). A marker gene from jellyfish, which produces **green fluorescent protein** (**GFP**) was used to confirm the transgenic status of a variety of ANDi's cells. ANDi is particularly significant in that he opens up the possibility of a near-human model organism for the evaluation of disease and therapy. He also brings the ethical questions surrounding transgenic research a little closer to the human situation, which some people are finding a little uncomfortable. It will certainly be interesting to follow the developments in transgenic technology over the next few years.

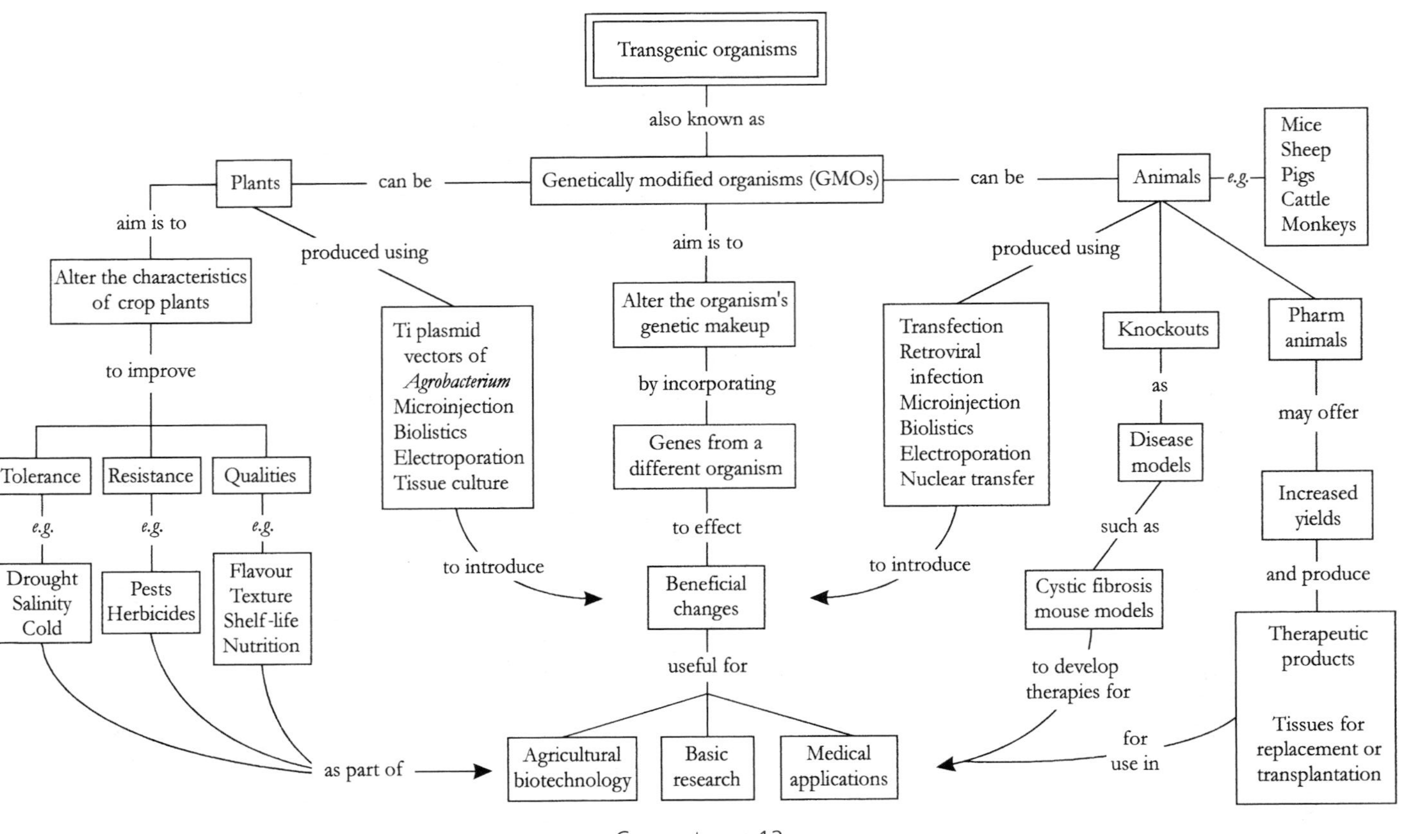

Concept map 12

13

The other sort of cloning

On 5th July 1996 a lamb was born at the Roslin Research Institute near Edinburgh. It was an apparently normal event, yet it marked the achievement of a milestone in biological science. The lamb was a clone, and was named Dolly. She was the first organism to be cloned from adult differentiated cells, which is what makes the achievement such a ground-breaking event. In this chapter we will look briefly at this area of genetic technology.

In this book so far, we have been considering the topic of **molecular cloning**, where the aim of an experimental process is to isolate a gene sequence for further analysis and use. In **organismal cloning**, the aim is to generate an organism from a cell that carries a complete set of genetic instructions. We have looked at the methods for generating transgenic organisms in Chapter 12, and a discussion of organismal cloning is a natural extension to this, although transgenic organisms are not necessarily (and at present are not usually) clones. In a similar way, a clone need not necessarily be transgenic. Thus, although not strictly part of gene manipulation technology, organismal cloning has become a major part of genetics in a broader sense. The public have latched on to cloning as an issue for concern, and thus a discussion of the topic is essential even in a book where the primary goal is to illustrate the techniques of gene manipulation.

13.1 Early thoughts and experiments

The announcement of the birth of Dolly in a paper in the journal *Nature* in February 1997 rocked the scientific community. However, the basic scientific

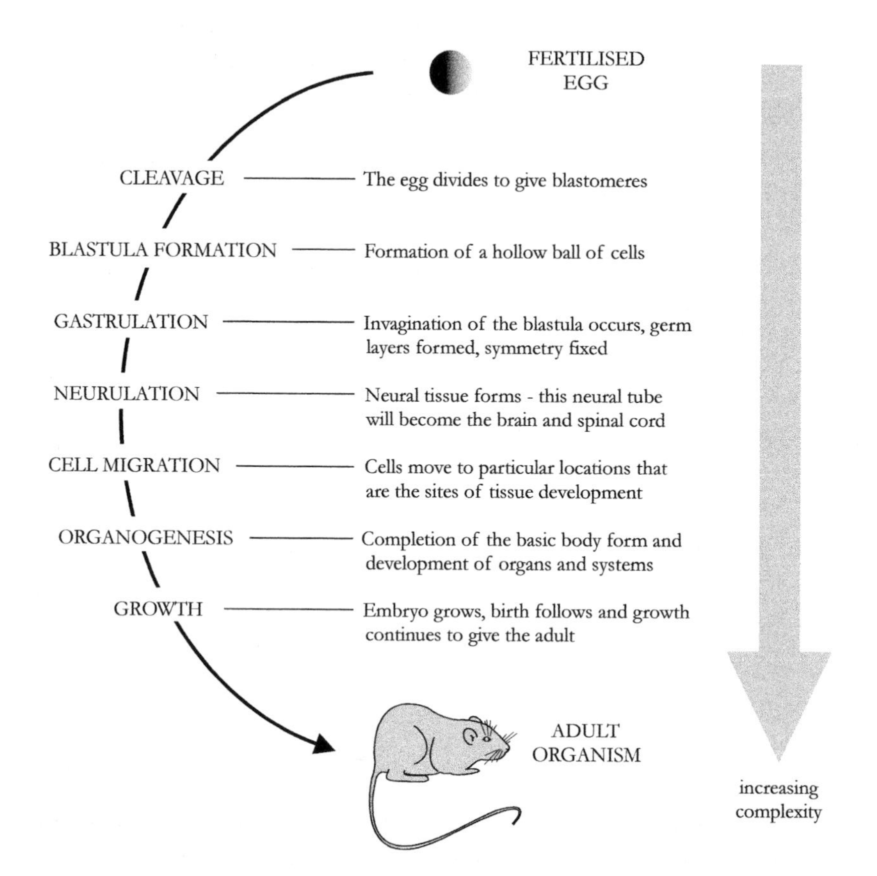

Fig. 13.1. Developmental sequence in the mouse. The journey from fertilised egg to adult organism involves cell division, cell differentiation, and the generation of complex patterns during embryogenesis. All these events must be coordinated and regulated in time and space if the process is to proceed successfully to completion. Modified after Nicholl (2000), *Cell and Molecular Biology*, Advanced Higher Monograph Series, Learning and Teaching Scotland.

principles that underpin organismal cloning have had a long history, and can be traced back to the early days of experimental embryology in the late 1800s and early 1900s. The early embryologists were seeking answers to the central question of development – how does a complex multicellular organism develop from a single fertilised egg?

There are two contrasting theories of development. One is that everything is preformed in some way, and development is simply the unfolding of this already existing pattern. This theory is called **preformationism,** and at its most extreme was thought of as a fully formed organism called a **homunculus** (a little man) sitting inside the sperm ready to 'grow' into a new individual during development. The alternative view is **epigenesis**, in which development is seen as an iterative process in which cells communicate with each other, and with their environment, as development proceeds. As with many opposing theories, there are aspects of each that can be considered valuable even today. Certainly, the genetic information is already 'pre-formed': thus the genome *could* be considered as a set of instructions that enable the unfolding of all structures in the embryo. However, with the growing appreciation of the importance of the proteins in the cell (the proteome), it became clear that development is indeed an interactive process that involves many factors, and that differential gene expression is the mechanism by which complexity is generated from the genetic information in the embryo. In addition to this control of gene expression, which essentially directs the process of embryogenesis, movement of cells and the formation of defined patterns enable structural complexity to arise. Overall, the process of development therefore involves a series of complex interactive steps in both a **temporal** and **spatial** context. This is illustrated in Fig. 13.1.

13.1.1 First steps towards cloning

Our current knowledge of embryological development, as shown in Fig. 13.1, has been established over a long period. The first 'embryologist' appears to have been Aristotle, who is credited with establishing an early version of the theory of epigenesis. More recently, August Weismann attempted to explain development as a unidirectional process. In 1885 he proposed that the genetic information of cells diminishes with each cell division as development proceeds. This set a number of scientists to work, trying to prove or disprove the theory. Results were somewhat contradictory, but in 1902 Hans Spemann managed to split a two-cell salamander embryo into two parts, using a hair taken from his baby son's head! Each half developed into a normal organism. Further work confirmed this result, and also showed that in cases where a nucleus was separated from the embryo, and cytoplasm was retained to effectively give a complete cell, a normal organism developed. Thus Weismann's idea of diminishing genetic resources was shown to be incorrect – all cells in developing embryos retained the ability to programme the entire course of development.

In 1938 Spemann published his book *Embryonic Development and Induction*, detailing his work. In this he proposed what he called 'a fantastical experiment', in which the nucleus would be removed from a cell and implanted into an egg from which the nucleus had been removed. Spemann could not see any way of achieving this, hence his caution by using the word 'fantastic'. However, he had proposed the technique that would later become known as cloning by **nuclear transfer**. Unfortunately, he did not live to see this attempted; the first cloning success was not achieved until 1952, 11 years after his death.

13.1.2 Nuclear totipotency

The work of Spemann was an important part of the development of modern embryology – indeed, he is often called the 'father' of this discipline. He had proposed nuclear transfer, and had demonstrated cloning by **embryo splitting**. The basis of these two techniques is shown in Fig. 13.2. The experiments with embryo splitting that had refuted Weismann's ideas showed that embryo cells retain the capacity to form all cell types. This became known as the concept of **nuclear totipotency**, which is now a fundamental part of developmental genetics. A cell is said to be totipotent if it can direct the formation of all cells in the organism. If it can direct a more limited number of cell types, it is said to be **pluripotent** or **multipotent**. Extending this along the developmental timeline, a cell that is not capable of directing development under appropriate conditions is said to be **irreversibly differentiated**.

Nuclear totipotency is in many ways self-evident, as an adult organism has many different types of cell. The original zygote genome, passed on by successive mitotic divisions, must have the capacity to generate these different cells. However, the key in developing cloning techniques was not that this idea was disputed, but rather centred around attempts to see when embryo cells became irreversibly differentiated, and perhaps lost the *capacity* (but not the *genes*) to be totipotent. The next experiments to shed light on this area were carried out in the 1950s.

13.2 Frogs and toads and carrots

Plant development is somewhat simpler than animal development, largely because there are fewer types of cell to arrange in the developing structure. However, the concept of nuclear totipotency is just as valid in plants as it is in animals. In fact, one of the early unequivocal experimental demonstrations of nuclear totipotency was provided by the humble carrot in the late 1950s. Work by

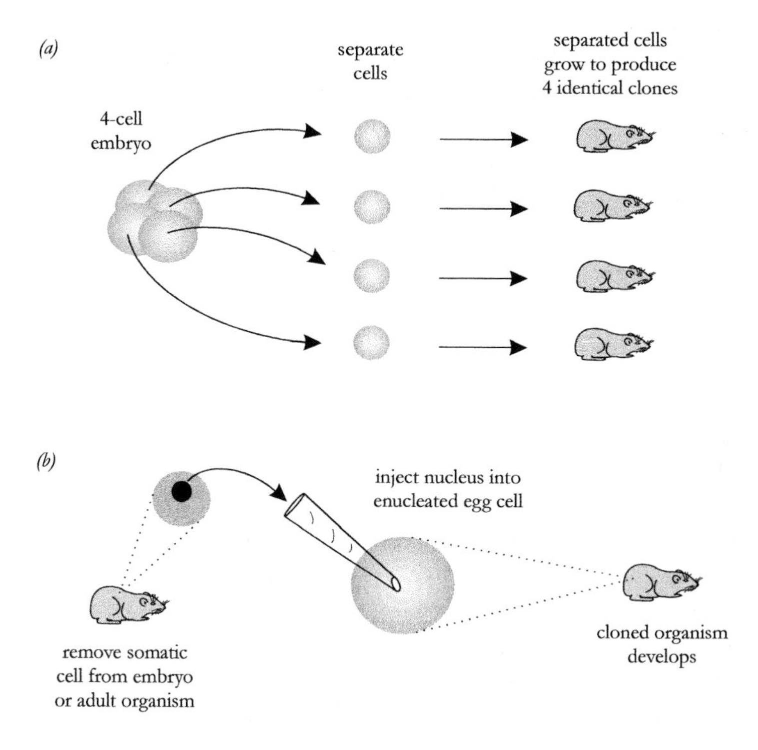

Fig. 13.2. Two methods for animal cloning. In (*a*) embryo splitting is shown. Cells from an early embryo (shown here as a four-cell embryo) are separated and allowed to continue development. Each cell directs the process of development to produce a new individual. The four organisms are genetically identical clones. In (*b*) the technique of nuclear transfer is shown. A nucleus from a somatic cell (from an embryo or an adult) is transplanted into an enucleated egg cell. If development can be sustained, a clone develops. Note that in this case the clone is not absolutely identical to its 'parent'. Mitochondrial DNA is inherited from the cytoplasm, and will therefore have been derived from the egg cell. This is known as a maternal inheritance pattern.

F.C. Steward and his colleagues at Cornell University showed that carrot plants could be regenerated from somatic (body) tissue, as shown in Fig. 13.3. This technique is now often used in the propagation of valuable plants in agriculture. The ability of plants to regenerate has of course been exploited for many years by taking cuttings and grafting – these are essentially asexual 'cloning' procedures.

Amphibians provide useful systems for embryological research in that the eggs are relatively large and plentiful, and development proceeds under less stringent environmental conditions than would be required for mammalian

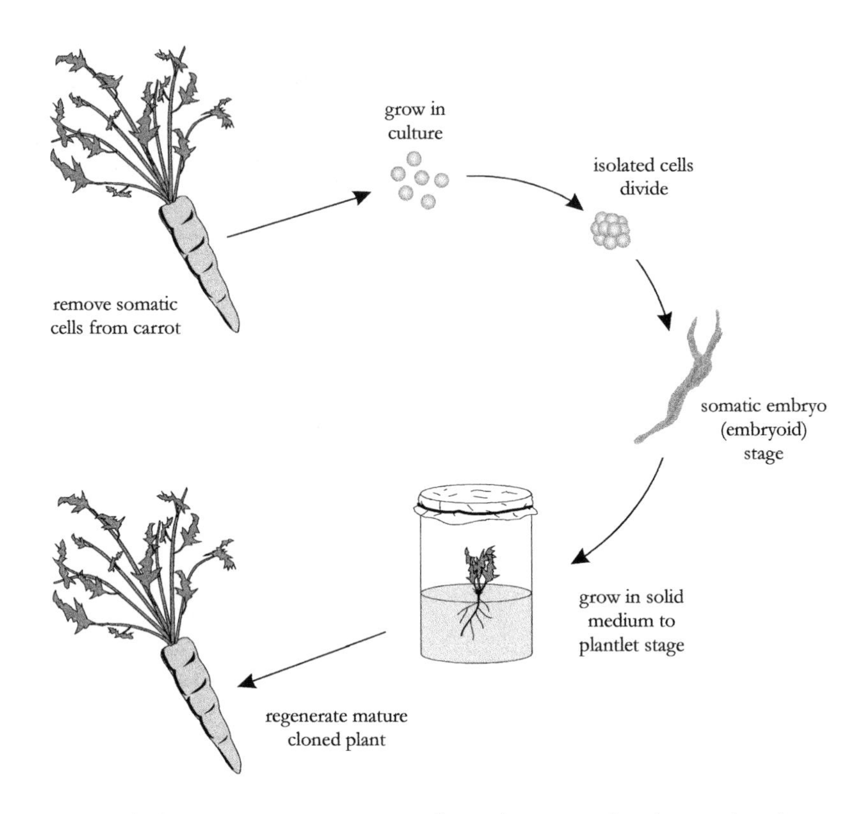

Fig. 13.3. Cloning of carrots. Somatic cells can be removed and grown in culture. Under appropriate conditions the cells begin to divide, then develop into somatic embryos known as embryoids. These can be transferred to a solid growth medium for plantlet development. The final stage is regeneration of the complete organism.

embryos. The frog *Rana pipiens* was used by Robert Briggs and Thomas King around 1952 to carry out Spemann's 'fantastical experiment'. Using nuclei isolated from blastula cells, they were able to generate cloned embryos, some of which developed into tadpoles (Fig. 13.4). As more work was done, it became apparent that early embryo cells could direct development, but that cell nuclei isolated from cells of older embryos were much less likely to generate clones. It was becoming clear that there was a point at which the cell DNA could not easily be 'deprogrammed' and used to direct the development of a new organism, and this remained the limiting factor in cloning research for many years. The work was later extended by John Gurdon at Oxford, using the toad *Xenopus laevis*. Some success was achieved, with fertile adult toads being generated from intestinal cell nuclei. In some of the experiments serial transfers were used, with the implanted nuclei allowed to develop, then these cells used

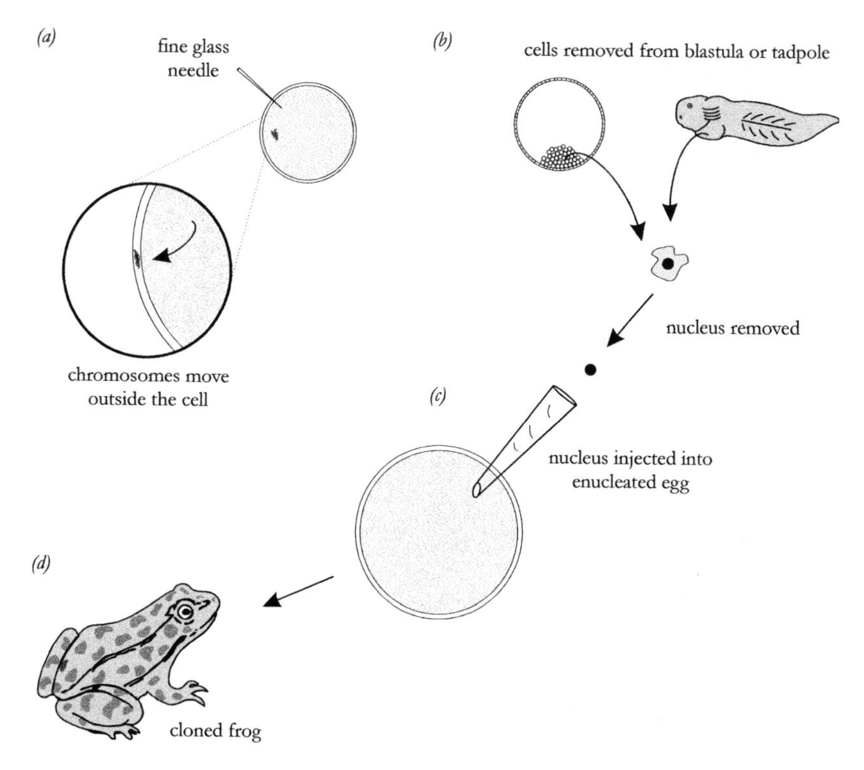

Fig. 13.4. Amphibian cloning. This is the type of procedure used by Briggs & King in 1952. (*a*) Egg cells are pricked with a fine glass needle, which activates the egg and causes the chromosomes to move outside the cytoplasm. Donor cell nuclei are removed from early embryos or tadpoles, as shown in (*b*). The donor nucleus is injected into the enucleated egg (*c*), and development can proceed. Early embryo donor nuclei gave better results than nuclei taken from later stages of development.

for the isolation of nuclei for further transfer. There was, however, a good deal of debate as to whether Gurdon had, in fact, used fully differentiated cells, or if contamination with primordial reproductive cells had produced the results.

13.3 A famous sheep – the breakthrough achieved

Despite the uncertainty surrounding Gurdon's experiments, the work with frogs and toads seemed to demonstrate that animal cloning by nuclear transfer was indeed a feasible proposition, if the right conditions could be established. Although the work seemed promising, cloning from a fully differentiated adult cell remained elusive. It was suspected that the manipulations used in the experiments could be damaging the donor nuclei, and by the 1970s it was clear that further development of the techniques was required. Reports of mice cloned from early embryo nuclei appeared in 1977, but the work was again somewhat inconclusive and difficult to repeat.

Development of the techniques for nuclear transfer cloning continued, with sheep and cattle as the main targets due to their potential for biotechnological applications in agriculture and in the production of therapeutic proteins. By the mid-1980s several groups had success with nuclear transfer from early embryos. A key figure at this time was Steen Willadsen, a Danish scientist who in 1984 achieved the first cloning of a sheep, using nuclear transfer from early embryo cells. In 1985, Willadsen cloned a cow from embryo cells, and in 1986 he achieved the same feat using cells from older embryos. The work with the older embryos (at the 64- to 128-cell stage) was not published, but Willadsen had demonstrated that cloning from older cells might not be impossible, as most people thought. By this stage several companies had become involved with cloning technology, particularly in cattle, and the future looked promising for the technology, the scientists who could do it, and the industry. However, as with the Flavr Savr transgenic tomato, there were problems in establishing cloning on a commercial footing, and by the early 1990s the promise had all but evaporated.

At around the same time that commercial interests were developing around cloning, a scientist called Ian Wilmut, working near Edinburgh in what would become the Roslin Institute, was busy generating transgenic sheep. He was keen to try to improve the efficiency of this somewhat hit-or-miss procedure, and cloning seemed an attractive way of doing this. If cells could be grown in culture, and the target transgenes added to these cells rather than being injected into fertilised eggs, the transgenic cells could be selected and used to clone the organism. This approach had been successful in mice, using embryonic stem cells, although it proved impossible to isolate the equivalent cells from sheep, cattle or pigs. However, older cells, derived from a foetus or adult organism, could be grown easily in culture. Thus a frustrating impasse existed – if only the older cells could be coaxed into directing development when used in a nuclear transfer experiment, then the process would work. This was the step that informed opinion said was impossible.

In 1986 Wilmut attended a scientific meeting in Dublin, where he heard about the 64- to 128-cell cattle cloning experiments from a vet who had worked with Willadsen. This encouraged him to continue with the cloning work. The key developments came when Keith Campbell joined Wilmut's group in 1990. Campbell was an expert on the cell cycle, and was able to develop techniques for growing cells in culture and then causing the cells to enter a quiescent stage of the cycle known as G_0. Wilmut and Campbell

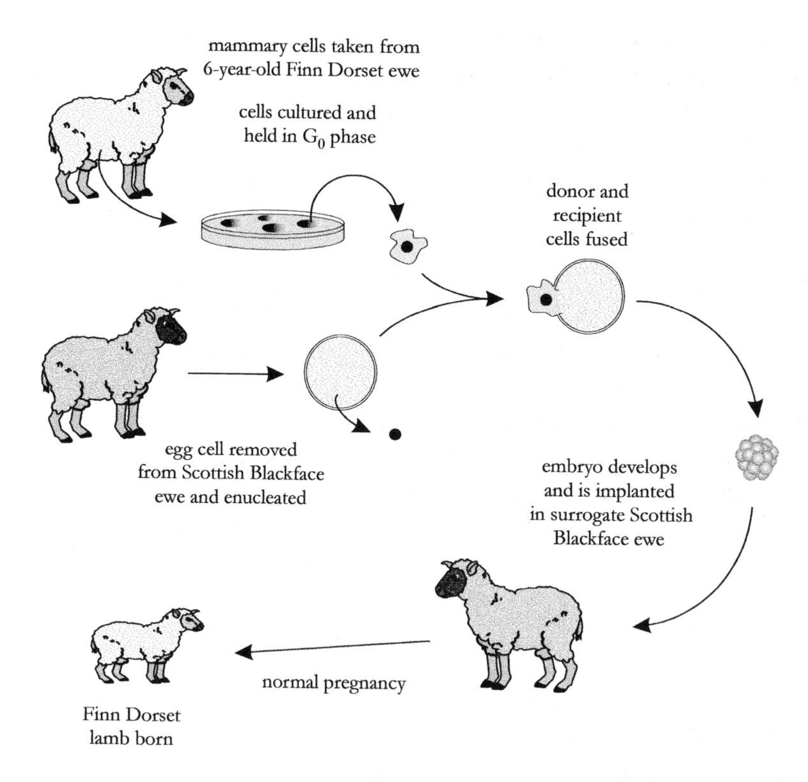

Fig. 13.5. The cloning method by which Dolly was produced. Mammary cells from an adult ewe were isolated, cultured and held in the G_0 phase of the cell cycle. These were the donor cells. Egg cells taken from a different breed were enucleated to act as recipients. The donor and recipient cells were fused and cultured. The embryos were then implanted into surrogate mothers and pregnancies established. Dolly was born some 5 months later.

thought that this might be a critical factor, perhaps the key to success. They were proved correct when, in 1995, Megan and Morag were born. These were lambs that had been produced by nuclear transfer using cultured cells derived from early embryos. Extensions to the work were planned, supported by PPL Therapeutics, the biotechnology firm set up in 1987 to commercialise transgenic sheep technology. Wilmut and Campbell devised a complex experiment in which they would use embryo cells, foetal cells and adult cells in a cloning procedure. A strange quirk of history appears at this point – the adult cells came from a vial that had been stored frozen at PPL for three years, and their source animal was long forgotten. However, it was known that the cells were from the udder of a 6-year-old Finn Dorset ewe. The cloning process is summarised in Fig. 13.5.

Fig. 13.6. Dolly – the first mammal to be cloned from a fully differentiated somatic adult cell. (Photograph courtesy of the Roslin Institute. Reproduced with permission.)

From the adult cell work, 29 embryos were produced from 277 udder cells. These were implanted into Scottish Blackface surrogate mothers, and some 148 days later, on 5th July 1996, one lamb was born. She was named Dolly, after the singer Dolly Parton (make the connection yourself!). Dolly is shown in Fig. 13.6. The impossible had been achieved.

13.4 Beyond Dolly

The birth of Dolly demonstrated that adult differentiated cells could, under the appropriate conditions, give rise to clones. Whilst the magnitude of this scientific achievement was appreciated by Wilmut and his colleagues at the time, the extent of the public reaction caught them a little by surprise. Suddenly Wilmut, Campbell and Ron James of PPL were in the limelight, unfamiliar ground for scientists. Even Dolly herself became something of a celebrity in the media. She was mated with a Welsh mountain ram in 1997 and gave birth to a female lamb called Bonnie in April 1998.

Having achieved success with Dolly, Wilmut and his colleagues went on to produce the first transgenic cloned sheep, a Poll Dorset clone (named Polly) carrying the gene for Factor IX. Thus the goal of producing transgenics using nuclear transfer technology had been achieved, and offers great potential for the future.

Cloning of mice, goats, cattle and monkeys has now been achieved, using either the embryo splitting technique or nuclear transfer. Cloning of human embryos by embryo splitting was reported as far back as 1993, although most scientists consider that nuclear transfer cloning of humans should not be attempted. However, proposals to clone humans have been put forward, notably by Richard Seed in 1997. At the time of writing, an Italian reproductive expert stated that he was setting out to establish cloning of humans. As with so many aspects of modern genetics, the future will, no doubt, hold many contentious and interesting developments in the field of organismal cloning. Regardless of how the application of cloning technology develops, when the history is written the central character will be a sheep named after a Country and Western singer – bizzare but strangely appropriate in a field that stirs the imagination like few others in science.

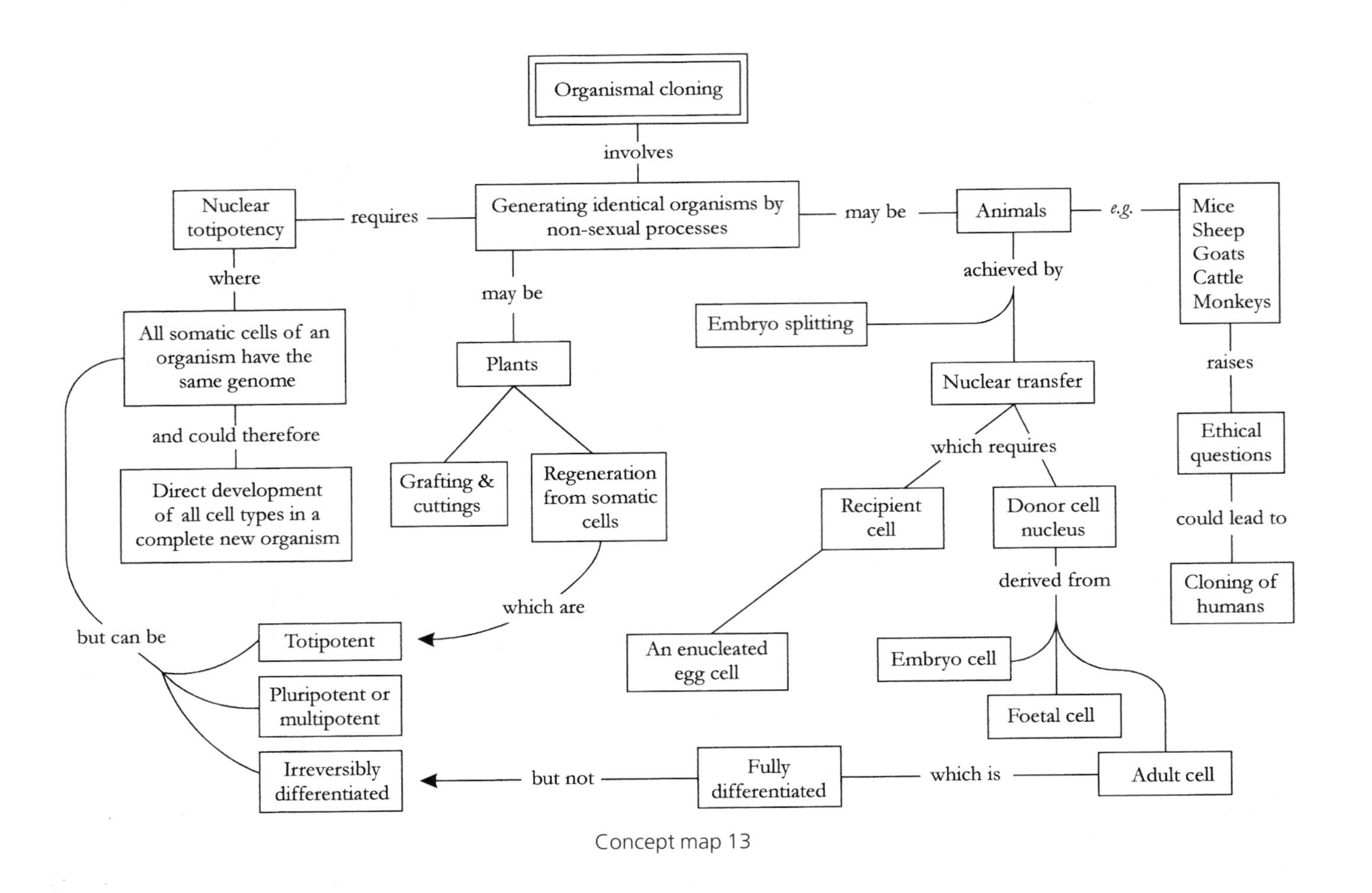

Concept map 13

14

Brave new world or genetic nightmare?

This final chapter is short. It does not answer any questions, but simply raises them for consideration. There are no 'correct' answers to these questions, as each must be addressed from the perspective of the individual, family, society, race or nation that is facing up to the situation. There are no diagrams or photographs, and very little factual information. However, the topics discussed are probably the most important that a student of genetic engineering can consider. In practical terms, relatively few people will ever go on to work in science and technology, but we will all have to cope with the consequences of gene-based research and its applications. Informed and vigorous debate is the only way that the developments of gene manipulation technology can become accepted and established.

14.1 Is science ethically and morally neutral?

It is often said that science *per se* is neither 'good' nor 'bad', and that it is therefore ethically and morally neutral. Whilst this may be true of science as a *process*, it is the developments and applications that arise from the scientific process that pose the ethical questions. The example that is often quoted is the development of the atomic bomb – the science was interesting and novel, and of itself ethically neutral, but the application (i.e. use of the devices in conflict) posed a completely different set of moral and ethical questions. Also, science is, of course, carried out by *scientists*, who are most definitely not ethically and morally neutral, as they demonstrate the same breadth and range of opinion as the rest of the human race. An assumption often made by the layman is that scientists and the scientific process are the same thing, which is unfortunate.

Despite the purist argument that science is in some way immune from ethical considerations, I believe that to separate the process from its applications is an artificial distinction. In the developed world we live in societies shaped by technology, which is derived from the application of scientific discoveries. We must all share the responsibility of policing the new genetic technology.

14.2 Elements of the ethics debate

Advances in the **basic science** of genetics usually pose few problems from an ethical standpoint. The major concerns are usually separate from the actual experiments – perhaps the use of animals in research, or the potential for transgenic crops to contaminate non-transgenic or wild populations. We will consider some of the ethical problems in medicine, biotechnology, transgenic organisms, and organismal cloning – thus following the arrangement of this book. However, there is considerable overlap in many of these areas, and it is again somewhat artificial to separate these topics.

In **medicine**, few would argue against the development of new drugs and therapies, where clear benefit is obtained. Perhaps the one area in the medically related applications of genetic research that is difficult is the human genome information. Genetic screening, and thus the possibility of genetic discrimination, is an area of active debate at the moment. The molecular diagnosis of genetically based disease is now well established, and the major ethical dilemmas tend to centre around whether or not a foetus should be aborted if a disease-causing trait is detected. If and when it becomes possible to screen routinely for polygenic and multifactorial traits, perhaps involving personality and predisposition to behavioural problems, the ethical picture will become even more complex. This whole area of predisposition, as opposed to a confirmed causal link between genes and disease, is a difficult area in which to establish any ethical rules, as many of the potential problems are as yet hypothetical.

The **biotechnology industry** is a difficult area to define, as the applications of gene technology in biotech applications are very diverse. The one ethical thread linking disparate applications is the influence of commercial interests. Patenting gene sequences raises questions, as does the production of products such as bovine somatotropin (BST). Many people see biotechnology applications as driven by commercial pressures, and some are uneasy with this.

Similar questions can be asked of any manufacturing process, but the use of biological material seems to set up a different attitude in many people. In some cases the apparent arrogance of biotechnology companies upsets people who might actually agree with the overall aims of the company, and several biotech companies have found that this public opinion can be a potent force in determining the success or failure of a product.

Transgenic organisms set up several ethical questions. The one thing that has been a little surprising is the reversal of the usual plant/animal debate as far as transgenesis is concerned. Traditionally, animal welfare has been the major source of difficulty between pressure groups, concerned individuals, scientists and regulators. Plants were largely ignored in the ethical debate until the late 1990s, when the public backlash against **genetically modified foods** began to influence what biotech companies were doing. Concerns were in two areas – the effect of GMO-derived foods on health, and the effect on the environment. The environmental debate in particular has been driven by many different groups, who claim that an ecological disaster might be waiting to emerge from GM plant technology due to cross-pollination. It is impossible to predict what might happen in such cases, although the protagonists of GM crops claim to have evaluated the risks. The simple answer is that we do not know what the long-term ecological effects might be.

Transgenic animals have not posed as big a problem as plants. This is a little surprising, but can be explained by the fact that animals are much easier to identify and contain (they need to copulate rather than pollinate!) and therefore any risk of transgenic traits getting into wild populations is much lower than for crop plants. Also, transgenic animals are often used in a medical context, where the benefits are obvious and generally appreciated. Welfare issues are still a concern, but the standard of care for animals in transgenic research is very high, and tightly monitored in most countries. Overall the acceptance of transgenic mouse models for disease, and transgenic animals as bioreactors, seems to pose less of a problem generally than aspects of plant biotechnology. The possibility of **xenotransplantation** offers hope but also raises questions, particularly for groups with specific religious objections to this type of application.

Organismal cloning is perhaps the most difficult area from an ethical viewpoint, as the possibility of human cloning exists and is being taken seriously by some. There is debate about the unique nature of personality, character, soul, nature or whatever it might be called; what would a cloned individual actually *be*? Views range from those who think that he or she would be in some sort of limbo, to those who see essentially no difference between a clone and a normal individual.

In a similar way to the progression of the Dolly research, and its extension into transgenic cloning (Polly), many people fear that the possibility of generating transgenic 'designer babies' may become reality at some point in the future. Normal conception, genetic screening, transgenesis to replace defective genes, implantation and development to full term? Or perhaps selection of a set of characteristics from a list, and production of the desired phenotype by manipulating the genome? This all sounds very fanciful at present, even absurd. In the early 1990s informed opinion said the same thing about cloning from adult cells. Dolly arrived in 1996.

14.1 Does Frankenstein's monster live inside Pandora's box?

The rather awful question that forms the title for this final section illustrates how two phrases that would not usually exist in one sentence can be joined together. This, of course, is exactly the essence of gene manipulation – splicing pieces of DNA together to generate recombinant molecules that would not exist in nature. In this book we have looked at the range of gene manipulation technology, from the basic techniques up to advanced applications. In trying to answer the question 'is it good or bad', we have seen that there can be no answer to this. As with any branch of human activity, the responsibility for using genetic technology lies with those who discover, adapt, implement and regulate it. However, the pressures that exist when commercial development of genetic engineering is undertaken can sometimes change the balance of responsibility. Most scientists ply their trade with honesty and integrity, and would not dream of falsifying results or inventing data. They take a special pride in what they do, and in a curious paradox remain detached from it, whilst being totally involved with it. Once the science becomes a technology, things are not quite so clear cut, and corporate responsibility is sometimes not quite so easy to define as individual responsibility.

In the few years since the first edition of this book was published, the developments in genetics and gene manipulation have been staggering. Even those working at the forefront of the various disciplines have often been surprised at how fast things have progressed, particularly with respect to genome sequencing projects. By the time this book is published, many more developments will have taken place. I hope that by reading this book you will be a little better prepared to assess these new developments, and take your part in the debate that we all must engage in to ensure that gene technology is used as a force for good.

Suggestions for further reading

A browse through a good academic bookshop or library is often the best way to spot texts that may be of interest. However, I have given a few suggestions that you may find useful. There are lots of different sources that can provide additional information, and this list is a very brief selection. I have not referred to the primary literature, as this is accessible from some of the texts and review journals listed below.

Books

The genetics texts listed below provide additional general information, and often extend the ideas presented in this book, or treat the topics with a different emphasis. All should be accessible to readers of this book. The more general texts are suitable as an introduction to the wider aspects of the subject – the stories behind the discoveries, the characters involved, and the ethical issues surrounding the work.

Genetics textbooks

Alcamo, I.E. (2000)
DNA Technology: The Awesome Skill 2nd edition
San Diego: Harcourt/Academic Press. ISBN 0120489201

Atherly, A.G., Girton, J.R. & McDonald, J.F. (1999)
The Science of Genetics
Orlando: Saunders College Publishing/Harcourt Brace. ISBN 0030332222

Brown, T.A. (1999)
Genomes
Oxford: Bios Scientific Publishers. ISBN 1859962017

Hartwell, L.H., Hood, L., Goldberg, M.L., Reynolds, A.E., Silver, L.M. & Veres, R.C. (2000)
Genetics: From Genes to Genomes
Boston: McGraw-Hill. ISBN 0075409232

Lewis, R. (2001)
Human Genetics: Concepts and Applications
Boston: McGraw-Hill. ISBN 0071180796

Sudbery, P. (1998)
Human Molecular Genetics
Harlow: Addison-Wesley Longman. ISBN 0582322669

General reading

Bishop, J. (1999)
Transgenic Mammals
Edinburgh: Pearson Education. ISBN 0582357306

Bruce, D. (1998)
Engineering Genesis: The Ethics of Genetic Engineering in Non-Human Species
London: Earthscan Publications. ISBN 1853835706

Burley, J. (1999) (editor)
The Genetic Revolution and Human Rights
Oxford: Oxford University Press. ISBN 0192862014

Cooper, D. K. C., Kemp, E., Reemtsma, K. & White, D. J. G. (1997)
Xenotransplantation: The Transplantation of Organs and Tissues Between Species 2nd edition
Berlin: Springer-Verlag. ISBN 3540617779

Kolata, G. (1997)
Clone: The Road to Dolly and the Path Ahead
London: Allen Lane, The Penguin Press. ISBN 0713992212

Wilmut, I., Campbell, K. & Tudge, C. (2000)
The Second Creation: The Age of Biological Control by the Scientists Who Cloned Dolly.
London: Headline Book Publishing. ISBN 0747275300

Review Journals

Review journals are a good source of detailed information about specific topics, and often provide extensive lists of relevant primary literature references. Journals may appear annually, monthly or weekly. Some examples are *Scientific American* and *New Scientist*, which publish readable articles on a wide range of topics. The *Annual Reviews* series offers volumes in *Biochemistry and Genetics*, and the *Trends* series publishes *Trends in Genetics*, *Biotechnology* and *Biochemical Sciences*. All should be available from good academic libraries and some large public libraries.

In addition to review journals, many front-line research journals also publish useful review articles and news about research fields. The best known are *Nature* (with versions such as *Nature Genetics* and *Nature Biotechnology*) and *Science*.

Using the World Wide Web

Over a very short timeframe, the **internet** and **world wide web** (**www**) have had an enormous impact on how we access and use information. The term *internet* is generally used to describe the network of computers that, together, provide the means to publish and share information, whilst the term *world wide web* is a more general description of the information that is published using the internet. However, the two terms are often used interchangeably, and the phrases 'surfing the net' and 'surfing the web' have become part of modern-day language.

If you are already a competent web user, you won't need this section. If you are a newcomer, I have provided some information to get you started. The web provides an exciting and easy to use medium that is both a goldmine and a minefield, so get to grips with the basics and have a go. It's fun!

Getting started

You need a computer, and a connection to the internet. Computer access to the web may be provided by your place of work or study, or you can set up a personal account from home. Universities, colleges, research institutes and companies will usually have sophisticated IT (information technology) resources, and often their own **intranet** system and **website**. Access to and from such sites is usually easy, fast and free from personal expense if you are a student or employee. Most institutions have rules about using the web from their computers, often called **acceptable use policies**. This is to prevent misuse of the system, such as downloading inappropriate material, accessing non-work-related chatrooms, playing games on the web, and so on.

To set up a personal account you need to have a computer with a **modem**, and you need to register with an **internet service provider** (**ISP**). Usually there are costs involved, and these may be for access and/or telephone charges. Most ISP companies offer a range of packages to suit your likely usage pattern, so it pays to examine the packages on offer to make sure that you get the best value. It is very easy to run up substantial time online, and if you are anything other than a very occasional user, a package with unmetered access is probably the best option to choose. Costs vary, and many ISP companies provide their own extensive website and news services, plus **e-mail** facilities. To actually look at web pages, you need to have a **web browser** on your computer. The two most common versions are *Netscape* and *Internet explorer*. One or other is usually supplied with your computer, or they can be obtained on CD ROM or downloaded from the web if you have access to another computer that already has an online connection.

Finding websites is generally straightforward if you know where you are going. Each website has an 'address' known as a **uniform resource locator** or **URL**. Some of these have been listed in the text already. A URL generally begins with **http://www.** followed by the specific address. Many recent textbooks have associated websites, as do research groups, university departments, companies, etc.

If you don't know the URL, or want to search for a particular type of website, your ISP will normally have its own **search engine**, and often enables access to several of these. A search engine enables you to look for information using a range of terms, and it is astonishing what you can find. However, caution is recommended....

Caution!

There is an awful lot of information on the web, and there is a lot of awful information there as well! It is very easy to get sidetracked and end up wasting a lot of time searching through sites that are of no value. I have just typed in some search terms using the search engine supplied by my personal ISP. The number of 'hits' for the various terms were as follows:

- genetics 354 863
- cloning 81 976
- DNA cloning 23 997
- sheep cloning 6 783
- cystic fibrosis 30 864

Two points are clear – firstly, the number of retrieved items is generally too large to be useful and, secondly, the more specific the search, the better. However, even with something like 'sheep cloning' there is still too much information to look through, so learning to use the advanced search facilities provided by various search engines is well worth the effort.

Types of website

The main point to be aware of when using the web is the there is very little restriction on **what** gets published, and **by whom**. The **peer review** system that is used in science is not often applied to web-based information, so not all information may be accurate. The type of website can often give some indication as to how reliable the information may be. The last part of the URL usually indicates the type of organisation, although this does vary from country to country. Examples are **.gov** for government site, **.ac.uk** for UK academic site (**.edu** used in some other countries), and **.com** or **.co** for commercial or company sites. The characteristics of these types of site can be described as follows:

- **government sites –** usually accurate information, but this can have a particular emphasis. Political influences and 'spin' can sometimes make certain types of information highly suspect for some government sites – a knowledge of the local situation is often useful.
- **research institutes and universities –** very often the most useful sites, as these tend to be run by people who are used to publishing information by the peer review mechanism, and therefore are aware of the need to be accurate and unbiased.
- **company sites –** again can provide a wealth of information, but this is obviously presented to support company policy in most cases. Many responsible companies take care to present a balanced view of any contentious issues, often including links to other sources of information.
- **publishers –** range from book publishers to online journals. Often provide excellent resource material, but for some journal sites access may require registration and payment of a fee for each article downloaded.
- **information 'gateway' sites –** many specialist research areas have associated sites that collate and distribute information about the latest developments. Often it is possible to register for e-mail updates to be sent directly when they are published. As with journals, some of these sites require registration and subscription to access the full range of facilities.
- **pressure group sites –** can provide a lot of useful information, very often gathered to counteract some particular claim made by another party. Despite the obvious dangers of accepting the views presented, many of these sites take great care to present accurate information. Can range from moderate to extreme in terms of the views presented.
- **personal sites –** prepared by individuals, therefore likely to be the most variable in terms of content and viewpoint. Range from highly respected sites by well-known experts to the sort of site that is best avoided as far as serious study is concerned.

The difficulty in weeding out the less useful (or downright weird) sites is that anyone with a flair for design can produce a website that looks impressive and authoritative. Conversely, some extremely useful and respected sites can look a little dull if a good web designer has not been involved. However, as long as you remember that critical evaluation is essential, you should have few problems.

Dive in and enjoy!

Part of the fun of using the web is that it *is* such a diverse and extensive medium, and it is constantly changing and expanding. Although I had contemplated providing a list of 'useful URLs', I decided against this for a number of reasons. Firstly, some URLs are given in particular parts of the text, and provide an entry point for particular topics. Secondly, if you are using this book as a course text, your tutor should be able to direct you to any specific websites that he or she uses to support the course. Thirdly, using a search engine is the most effective way of getting up-to-date information about what is available, and by finding a few relevant sites, you can quickly build up a set of links that take you into other related sites. Good surfing!

Glossary

Abundance class Refers to the relative abundance of different mRNA molecules in a cell at any given time.

ADA Adenosine deaminase, deficiency results in SCIDS (q.v.)

Adaptor A synthetic single-stranded non self-complementary oligonucleotide used in conjunction with a linker to add cohesive ends to DNA molecules.

Adenine (A) Nitrogenous base found in DNA and RNA.

Adeno-associated virus Virus used in gene therapy delivery methods.

Adenovirus Virus that can infect through nasal passages, used in gene therapy delivery methods.

Aetiology Of disease; relating to the causes of the disease.

Agrobacterium tumefaciens Bacterium that infects plants and causes crown gall disease (q.v.). Carries a plasmid (the Ti plasmid) used for gene manipulation in plants.

Agarose Jelly-like matrix, extracted from seaweed, used as a support in the separation of nucleic acids by gel electrophoresis.

Alkaline phosphatase An enzyme that removes 5′ phosphate groups from the ends of DNA molecules, leaving 5′ hydroxyl groups.

Allele One of two or more variants of a particular gene.

Allele-specific oligonucleotide Oligonucleotide with a sequence that can be matched precisely to a particular allele by using stringent hybridisation conditions.

Alpha-peptide Part of the β-galactosidase protein, encoded by the *lacZ′* gene fragment.

Ampicillin (Ap) A semisynthetic β-lactam antibiotic.

Aneuploidy Variation in chromosome number where single chromosomes are affected, thus the chromosome complement is not an exact multiple of the haploid chromosome number.

Animal model Usually a transgenic mouse in which a disease state has been engineered. See also *knockout mouse*, *knockin mouse*, *oncomouse*.

Antibody An immunoglobulin that specifically recognises and binds to an antigenic determinant on an antigen.

Anticodon The three bases on a tRNA molecule that are complementary to the codon on the mRNA.

Antigen A molecule that is bound by an antibody. Also used to describe molecules that can induce an immune response, although these are more properly described as immunogens.

Antiparallel The arrangement of complementary DNA strands, which run in different directions with respect to their 5′ 3′ polarity.

Antisense RNA Produced from a gene sequence inserted in the opposite orientation, so that the transcript is complementary to the normal mRNA and can therefore bind to it and prevent translation.

Arabidopsis thaliana Small plant favoured as a research organism for plant molecular biologists.

Arbitrarily primed PCR PCR using random primers, useful in the technique of RAPD analysis (q.v.).

ASO See allele-specific oligonucleotide.

Autoradiograph Image produced on X-ray film in response to the emission of radioactive particles.

Autosome A chromosome that is not a sex chromosome.

Auxotroph A cell that requires nutritional supplements for growth.

Bacillus thuringiensis Bacterium used in crop protection, and in the generation of Bt plants that are resistant to insect attack. The bacterium produces a toxin that affects the insect.

Bacteriophage A bacterial virus.

Baculovirus A particular type of virus that infects insect cells, producing large inclusions in the infected cells.

Bal 31 nuclease An exonuclease that degrades both strands of a DNA molecule at the same time.

Bacterial alkaline phosphatase (BAP) See *alkaline phosphatase.*

Bioinformatics The emerging discipline of collating and analysing biological information, especially genome sequence information.

Biolistic Refers to a method of introducing DNA into cells by bombarding them with microprojectiles, which carry the DNA.

Blunt ends DNA termini without overhanging 3′ or 5′ ends. Also known as *flush ends.*

Bovine somatotropin (BST) Bovine growth hormone, produced as rBST for use in dairy cattle to increase milk production.

Bt plants Plants which carry the toxin-producing gene from *Bacillus thuringiensis* as a means to protect the plant from insect attack.

C terminus Carboxyl terminus, defined by the –COOH group of an amino acid or protein.

CAAT box A sequence located approximately 75 base-pairs upstream from eukaryotic transcription start sites. This sequence is one of those that enhance binding of RNA polymerase.

Caenorhabditis elegans A nematode worm used as a model organism in developmental and molecular studies.

Calf intestinal phosphatase (CIP) See *alkaline phosphatase*.

Cap A chemical modification that is added to the 5′ end of a eukaryotic mRNA molecule during post-transcriptional processing of the primary transcript.

Capsid The protein coat of a virus.

cDNA DNA that is made by copying mRNA using the enzyme reverse transcriptase.

cDNA library A collection of clones prepared from the mRNA of a given cell or tissue type, representing the genetic information expressed by such cells.

Central dogma Statement regarding the unidirectional transfer of information from DNA to RNA to protein.

CFTR gene (protein) Cystic fibrosis transmembrane conductance regulator, the gene and protein involved in defective ion transport that causes cystic fibrosis.

Chimaera An organism (usually transgenic) composed of cells with different genotypes.

Chromosome A DNA molecule carrying a set of genes. There may be a single chromosome, as in bacteria, or multiple chromosomes, as in eukaryotic organisms.

Chromosome jumping Technique used to isolate non-contiguous regions of DNA by 'jumping' across gaps that may appear as a consequence of uncloned regions of DNA in a gene library.

Chromosome walking Technique used to isolate contiguous cloned DNA fragments by using each fragment as a probe to isolate adjacent cloned regions.

Chymosin (chymase) Enzyme used in cheese production, available as recombinant product.

***Cis*-acting element** A DNA sequence that exerts its effect only when on the same DNA molecule as the sequence it acts on. For example, the CAAT box (q.v.) is a *cis*-acting element for transcription in eukaryotes.

Cistron A sequence of bases in DNA that specifies one polypeptide.

Clone (1) A colony of identical organisms; often used to describe a cell carrying a recombinant DNA fragment. (2) Used as a verb to describe the generation of recombinants. (3) A complex organism (e.g. sheep) generated from a totipotent cell nucleus by nuclear transfer into an enucleated ovum.

Clone bank See *cDNA library*, *genomic library*.

Codon The three bases in mRNA that specify a particular amino acid during translation.

Cohesive ends Those ends (termini) of DNA molecules that have short complementary sequences that can stick together to join two DNA molecules. Often generated by restriction enzymes.

Competent Refers to bacterial cells that are able to take up exogenous DNA.

Competitor RT-PCR Technique used to quantify the amount of PCR product by spiking samples with known amounts of a competitor sequence.

Complementary DNA See *cDNA*.

Complementation Process by which genes on different DNA molecules interact. Usually a protein product is involved, as this is a diffusible molecule that can exert its effect away from the DNA itself. For example, a *lacZ*$^+$ gene on a plasmid can complement a mutant (*lacZ*$^-$) gene on the chromosome by enabling the synthesis of β-galactosidase.

Concatemer A DNA molecule composed of a number of individual pieces joined together via cohesive ends (q.v.).

Congenital Present at birth, usually used to describe genetically derived abnormalities.

Conjugation Plasmid-mediated transfer of genetic material from a 'male' donor bacterium to a 'female' recipient.

Consensus sequence A sequence that is found in most examples of a particular genetic element, and which shows a high degree of conservation. An example is the CAAT box (q.v.).

Copy number (1) The number of plasmid molecules in a bacterial cell. (2) The number of copies of a gene in the genome of an organism.

***cos* site** The region generated when the cohesive ends of λ DNA join together.

Cosmid A hybrid vector made up of plasmid sequences and the cohesive ends (*cos* sites) of bacteriophage lambda.

Crown gall disease Plant disease caused by the Ti plasmid of *Agrobacterium tumefaciens*, in which a 'crown gall' of tissue is produced after infection.

Cyanogen bromide Chemical used to cleave a fusion protein product from the N-terminal vector-encoded sequence after synthesis.

Cystic fibrosis Disease affecting lungs and other tissues, caused by ion transport defects in the CFTR gene (q.v.).

Cytosine (C) Nitrogenous base found in DNA and RNA.

Deletion Change to the genetic material caused by removal of part of the sequence of bases in DNA.

Deoxynucleoside triphosphate (dNTP) Triphosphorylated ('high energy') precursor required for synthesis of DNA, where N refers to one of the four bases (A,G,T or C).

Deoxyribonucleic acid (DNA) A condensation heteropolymer composed of nucleotides. DNA is the primary genetic material in all organisms apart from some RNA viruses. Usually double-stranded.

Deoxyribose The sugar found in DNA.

Deoxyribonuclease (DNase) An nuclease enzyme that hydrolyses (degrades) single- and double-stranded DNA.

Dideoxynucleoside triphosphate (ddNTP) A modified form of dNTP used as a chain terminator in DNA sequencing.

Diploid Having two sets of chromosomes. Cf. *haploid.*

Disarmed vector A vector in which some characteristic (e.g. conjugation) has been disabled.

DMD See *Duchenne muscular dystrophy.*

DNA chip A DNA microarray used in the analysis of gene structure and expression. Consists of oligonucleotide sequences immobilised on a 'chip' array.

DNA fingerprinting See *genetic fingerprinting.*

DNA footprinting Method of identifying regions of DNA to which regulatory proteins will bind.

DNA ligase Enzyme used for joining DNA molecules by the formation of a phosphodiester bond between a 5′ phosphate and a 3′ OH group.

DNA microarray See *DNA chip.*

DNA polymerase An enzyme that synthesises a copy of a DNA template.

DNA profiling Term used to describe the various methods for analysing DNA to establish identity of an individual.

Dominant An allele that is expressed and appears in the phenotype in heterozygous individuals. Cf. *recessive.*

Dot-blot Technique in which small spots, or 'dots', of nucleic acid are immobilised on a nitrocellulose or nylon membrane for hybridisation.

Downstream processing Refers to the procedures used to purify products (usually proteins) after they have been expressed in bacterial, fungal or mammalian cells.

Drosophila melanogaster Fruit fly used as a model organism in genetic, developmental and molecular studies.

Duchenne muscular dystrophy X-linked (q.v.) muscle-wasting disease caused by defects in the gene for the protein dystrophin (q.v.).

Dystrophin Large protein linking the cytoskeleton to the muscle cell membrane, defects in which cause muscular dystrophy.

Electroporation Technique for introducing DNA into cells by giving a transient electric pulse.

ELSI Sometimes used as shorthand to describe the ethical, legal and social implications of genetic engineering.

Embryo splitting Technique used to clone organisms by separating cells in the early embryo, which then go on to direct development and produce identical copies of the organism.

End labelling Adding a radioactive molecule onto the end(s) of a polynucleotide.

Endonuclease An enzyme that cuts within a nucleic acid molecule, as opposed to an exonuclease (q.v.), which digests DNA from one or both ends.

Enhancer A sequence that enhances transcription from the promoter of a eukaryotic gene. May be several thousand base-pairs away from the promoter.

Enzyme A protein that catalyses a specific reaction.

Enzyme replacement therapy Therapeutic procedure in which a defective enzyme function is restored by replacing the enzyme itself. Cf. *gene therapy.*

Epigenesis Theory of development that regards the process as an iterative series of steps, in which the various signals and control events interact to regulate development.

Escherichia coli The most commonly used bacterium in molecular biology.

Ethidium bromide A molecule that binds to DNA and fluoresces when viewed under ultraviolet light. Used as a stain for DNA.

Eukaryotic The property of having a membrane-bound nucleus.

Exon Region of a eukaryotic gene that is expressed *via* mRNA.

Exonuclease An enzyme that digests a nucleic acid molecule from one or both ends.

Expressivity The degree to which a particular genotype generates its effect in the phenotype. Cf. *penetrance.*

Extrachromosomal element A DNA molecule that is not part of the host cell chromosome.

Ex vivo Outside the body. Usually used to describe gene therapy procedure in which the manipulations are performed outside the body, and the altered cells returned after processing. Cf. *in vivo*, *in vitro.*

Flavr Savr (*sic*) Transgenic tomato in which polygalacturonase (q.v.) synthesis is restricted using antisense technology. Despite the novel science, the Flavr Savr was not a commercial success.

Flush ends See *blunt ends.*

Foldback DNA Class of DNA which has palindromic or inverted repeat regions that re-anneal rapidly when duplex DNA is denatured.

Fusion protein A hybrid recombinant protein that contains vector-encoded amino acid residues at the N terminus.

ß-Galactosidase An enzyme encoded by the *lacZ* gene. Splits lactose into glucose and galactose.

Gamete Refers to the haploid male (sperm) and female (egg) cells that fuse to produce the diploid zygote (q.v.) during sexual reproduction.

Gel electrophoresis Technique for separating nucleic acid molecules on the basis of their movement through a gel matrix under the influence of an electric field. See *agarose* and *polyacrylamide.*

Gel retardation Method of determining protein-binding sites on DNA fragments on the basis of their reduced mobility, relative to unbound DNA, in gel electrophoresis experiments.

Gene The unit of inheritance, located on a chromosome. In molecular terms, usually taken to mean a region of DNA that encodes one function. Broadly, therefore, one gene encodes one protein.

Gene bank See *genomic library.*

Gene cloning The isolation of individual genes by generating recombinant DNA molecules, which are then propagated in a host cell which produces a clone that contains a single fragment of the target DNA.

Gene protection technology Range of techniques used to ensure that particular commercially derived recombinant constructs cannot be used without some sort of control or process, usually supplied by the company marketing the recombinant. Also known as genetic use restriction technology and genetic trait control technology.

Gene therapy The use of cloned genes in the treatment of genetically derived malfunctions. May be delivered *in vivo* or *ex vivo*. May be offered as gene addition or gene replacement versions.

Genetic code The triplet codons that determine the types of amino acid that are inserted into a polypeptide during translation. There are 61 codons for 20 amino acids (plus three stop codons), and the code is therefore referred to as *degenerate*.

Genetic fingerprinting A method which uses radioactive probes to identify bands derived from hypervariable regions of DNA (q.v.). The band pattern is unique for an individual, and can be used to establish identity or family relationships.

Genetic mapping Low-resolution method to assign gene locations (loci) to their position on the chromosome. Cf. *recombination frequency mapping, physical mapping.*

Genetic marker A phenotypic characteristic that can be ascribed to a particular gene.

Genetic trait control technology Version of gene protection technology (q.v.), sometimes called 'traitor technology'.

Genetic use restriction technology (GURT) See *gene protection technology*

Genetically modified organism (GMO) An organism in which a genetic change has been engineered. Usually used to describe transgenic plants and animals.

Genome Used to describe the complete genetic complement of a virus, cell or organism.

Genomics The study of genomes, particularly genome sequencing.

Genomic library A collection of clones which together represent the entire genome of an organism.

Genotype The genetic constitution of an organism. Cf. *phenotype.*

Germ line Gamete producing (reproductive) cells that give rise to eggs and sperm.

GMO See *genetically modified organism.*

Guanine (G) Nitrogenous base found in DNA and RNA.

GURT See *genetic use restriction technology, gene protection technology.*

Haploid Having one set of chromosomes. Cf. *diploid.*

Heterologous Refers to gene sequences that are not identical, but show variable degrees of similarity.

Heteropolymer A polymer composed of different types of monomer. Most protein and nucleic acid molecules are heteropolymers.

Heterozygous Refers to a diploid organism (cell or nucleus) which has two different alleles at a particular locus.

Hogness box See *TATA box.*

Homologous (1) Refers to paired chromosomes in diploid organisms. (2) Used to strictly describe DNA sequences that are identical; however, the percentage homology between related sequences is sometimes quoted.

Homopolymer A polymer composed of only one type of monomer, such as polyphenylalanine (protein) or polyadenine (nucleic acid).

Homozygous Refers to a diploid organism (cell or nucleus) which has identical alleles at a particular locus.

Host A cell used to propagate recombinant DNA molecules.

Hybrid-arrest translation Techniques used to identify the protein product of a cloned gene, in which translation of its mRNA is prevented by the formation of a DNA·mRNA hybrid.

Hybrid-release translation Technique in which a particular mRNA is selected by hybridisation with its homologous cloned DNA sequence, and is then translated to give a protein product that can be identified.

Hybridisation The joining together of artificially separated nucleic acid molecules *via* hydrogen bonding between complementary bases.

Hyperchromic effect Change in absorbance of nucleic acids, depending on the relative amounts of single-stranded and double-stranded forms. Used as a measurement in denaturation/renaturation studies.

Hypervariable region (HVR) A region in a genome that is composed of a variable number of repeated sequences and is diagnostic for the individual. See *genetic fingerprinting.*

Ice-minus bacteria Bacteria engineered to disrupt the normal ice-forming process, used to protect plants from frost damage.

IGF-1 See *insulin-like growth factor.*

Insertion vector A bacteriophage vector that has a single cloning site into which DNA is inserted.

Insulin-like growth factor (IGF-1) Polypeptide hormone, synthesis of which is stimulated by growth hormone. Implicated in some concerns about the safety of using recombinant bovine growth hormone in cattle to increase milk yields.

Intervening sequence Region in a eukaryotic gene that is not expressed *via* the processed mRNA.

Intron See *intervening sequence.*

Inverted repeat A short sequence of DNA that is repeated, usually at the ends of a longer sequence, in a reverse orientation.

In vitro Literally ‘in glass’, meaning in the test-tube, rather than in the cell or organism.

In vivo Literally ‘in life’, meaning the natural situation, within a cell or organism.

IPTG *iso*-Propyl-thiogalactoside, a gratuitous inducer which de-represses transcription of the *lac* operon.

Kilobase (kb) 10^3 bases or base-pairs, used as a unit for measuring or specifying the length of DNA or RNA molecules.

Klenow fragment A fragment of DNA polymerase I that lacks the $5'$ $3'$ exonuclease activity.

Knockin mouse A transgenic mouse in which a gene function has been added or 'knocked in'. Used primarily to generate animal models for the study of human disease. Cf. *knockout mouse.*

Knockout mouse A transgenic mouse in which a gene function has been disrupted or 'knocked out'. Used primarily to generate animal models for the study of human disease, e.g cystic fibrosis. Cf. *knockin mouse.*

Linkage mapping Genetic mapping (q.v.) technique used to establish the degree of linkage between genes. See also *recombination frequency mapping.*

Linker A synthetic self-complementary oligonucleotide that contains a restriction enzyme recognition site. Used to add cohesive ends (q.v.) to DNA molecules that have blunt ends (q.v.).

Lipase Enzyme that hydrolyses fats (lipids).

Liposome (lipoplex) Lipid-based method for delivering gene therapy.

Locus The site at which a gene is located on a chromosome.

Lysogenic Refers to bacteriophage infection that does not cause lysis of the host cell.

Lytic Refers to bacteriophage infection that causes lysis of the host cell.

Maternal inheritance Pattern of inheritance from female cytoplasm. Mitochondrial genes are inherited in this way, as the mitochondria are inherited with the ovum.

Mega (M) SI prefix, 10^6.

Messenger RNA (mRNA) The ribonucleic acid molecule transcribed from DNA that carries the codons specifying the sequence of amino acids in a protein.

Micro (μ) SI prefix, 10^{-6}.

Microinjection Introduction of DNA into the nucleus or cytoplasm of a cell by insertion of a microcapillary and direct injection.

Microsatellite DNA Type of sequence repeated many times in the genome. Based on dinucleotide repeats, microsatellites are highly variable and can be used in mapping and profiling studies.

Milli (m) SI prefix, 10^{-3}.

Minisatellite DNA Type of sequence based on variable number tandem repeats (VNTRs; q.v.). Used in genetic mapping and profiling studies.

Molecular cloning Alternative term for gene cloning.

Molecular ecology Use of molecular biology and recombinant DNA techniques in studying ecological topics.

Molecular paleontology Use of molecular techniques to investigate the past, as in DNA profiling from mummified or fossilised samples.

Monocistronic Refers to an RNA molecule encoding one function.

Monogenic Trait caused by a single gene. Cf. *polygenic.*

Monomer The unit that makes up a polymer. Nucleotides and amino acids are the monomers for nucleic acids and proteins, respectively.

Monosomic Diploid cells in which one of a homologous pair of chromosomes has been lost. Cf. *trisomy.*

Monozygotic Refers to identical twins, generated from the splitting of a single embryo at an early stage.
Mosaic An embryo or organism in which not all the cells carry identical genomes.
Multifactorial Caused by many factors, e.g. genetic trait in which many genes and environmental influences may be involved.
Multi-locus probe DNA probe used to identify several bands in a DNA fingerprint or profile. Generates the 'bar code' pattern in a genetic fingerprint.
Multiple cloning site (MCS) A short region of DNA in a vector that has recognition sites for several restriction enzymes.
Multipotent Cell which can give rise to a range of differentiated cells. Cf. *totipotent*, *pluripotent*.
Mutagenesis The process of inducing mutations in DNA.
Mutant An organism (or gene) carrying a genetic mutation.
Mutation An alteration to the sequence of bases in DNA. May be caused by insertion, deletion or modification of bases.

Nano (n) SI prefix, 10^{-9}.
Native protein A recombinant protein that is synthesised from its own N terminus, rather than from an N terminus supplied by the cloning vector.
Nested fragments A series of nucleic acid fragments that differ from each other (in terms of length) by one or only a few nucleotides.
Nick translation Method for labelling DNA with radioactive dNTPs.
Northern blotting Transfer of RNA molecules onto membranes for the detection of specific sequences by hybridisation.
N terminus Amino terminus, defined by the $-NH_2$ group of an amino acid or protein.
Nuclear transfer Method for cloning organisms in which a donor nucleus is taken from a somatic cell and transferred to the recipient ovum.
Nuclease An enzyme that hydrolyses phosphodiester bonds.
Nucleoside A nitrogenous base bound to a sugar.
Nucleotide A nucleoside bound to a phosphate group.
Nucleoid Region of a bacterial cell in which the genetic material is located.
Nucleus Membrane-bound region in a eukaryotic cell that contains the genetic material.

Oligo Prefix meaning few, as in oligonucleotide or oligopeptide.
Oligo(dT)-cellulose Short sequence of deoxythymidine residues linked to a cellulose matrix, used in the purification of eukaryotic mRNA.
Oligolabelling See *primer extension*.
Oligomer General term for a short sequence of monomers.
Oligonucleotide A short sequence of nucleotides.
Oligonucleotide-directed mutagenesis Process by which a defined alteration is made to DNA using a synthetic oligonucleotide.

Oncomouse Transgenic mouse engineered to be susceptible to cancer.

Oocyte Stage in development of the female gamete or ovum (egg). Often the terms oocyte and ovum are used interchangeably.

Operator Region of an operon, close to the promoter, to which a repressor protein binds.

Operon A cluster of bacterial genes under the control of a single regulatory region.

Organismal cloning The production of an identical copy of an individual organism by techniques such as embryo splitting or nuclear transfer. Used to distinguish the process from molecular cloning (q.v.).

Ovum The mature female gamete or egg cell, derived from the oocyte. Often the terms ovum and oocyte are used interchangeably.

Palindrome A DNA sequence that reads the same on both strands when read in the same (e.g. 5′ 3′) direction. Examples include many restriction enzyme recognition sites.

Pedigree analysis Determination of the transmission characteristics of a particular gene by examination of family histories.

Penetrance The proportion of individuals with a particular genotype that show the genotypic characteristic in the phenotype. Cf. *expressivity*.

PCR See *polymerase chain reaction*.

Phage See *bacteriophage*.

Phagemid A vector containing plasmid and phage sequences.

Pharm animal Transgenic animal used for the production of pharmaceuticals.

Phenotype The observable characteristics of an organism, determined both by its genotype (q.v.) and its environment.

Phosphodiester bond A bond formed between the 5′ phosphate and the 3′ hydroxyl groups of two nucleotides.

Physical mapping Mapping genes with reference to their physical location on the chromosome. Generates the next level of detail compared to genetic mapping (q.v.).

Physical marker A sequence-based tag that labels a region of the genome. There are several such tags that can be used in mapping studies. Cf. *RFLP*, *STS*.

Pico (p) SI prefix, 10^{-12}.

Plaque A cleared area on a bacterial lawn caused by infection by a lytic bacteriophage.

Plasmid A circular extrachromosomal element found naturally in bacteria and some other organisms. Engineered plasmids are used extensively as vectors for cloning.

Ploidy number Refers to the number of sets of chromosomes, e.g. haploid, diploid, triploid, etc.

Pluripotent Cell which can give rise to a range of differentiated cells. Cf. *multipotent*, *totipotent*.

Polyacrylamide A cross-linked matrix for gel electrophoresis (q.v.) of small fragments of nucleic acids, primarily used for electrophoresis of DNA. Also used for electrophoresis of proteins.

Polyadenylic acid A string of adenine residues. Poly(A) tails are found at the 3′ ends of most eukaryotic mRNA molecules.

Polycistronic Refers to an RNA molecule encoding more that one function. Many bacterial operons are expressed *via* polycistronic mRNAs.

Polygalacturonase Enzyme involved in pectin degradation. Target for antisense control in the Flavr Savr tomato (q.v.).

Polygenic trait A trait determined by the interaction of more than one gene, e.g. eye colour in humans.

Polyhedra Capsid structures in baculoviruses, composed of the protein polyhedrin.

Polylinker See *multiple cloning site.*

Polymer A long sequence of monomers.

Polymerase An enzyme that synthesises a copy of a nucleic acid.

Polymerase chain reaction (PCR) A method for the selective amplification of DNA sequences. Several variants exist for different applications.

Polymorphism Refers to the occurrence of many allelic variants of a particular gene or DNA sequence motif. Can be used to identify individuals by genetic mapping and DNA profiling techniques.

Polynucleotide A polymer made up of nucleotide monomers.

Polynucleotide kinase (PNK) An enzyme that catalyses the transfer of a phosphate group onto a 5′ hydroxyl group.

Polypeptide A chain of amino acid residues. Cf. *protein.*

Polystuffer An expendable stuffer fragment in a vector that is composed of many repeated sequences.

Positional cloning Cloning genes for which little information is available apart from their location on the chromosome.

Preformationism Refers to the idea that all development is pre-coded in the zygote, and that development is simply the unfolding of this information. Now considered too simplistic. Cf. *epigenesis.*

Pribnow box Sequence found in prokaryotic promoters that is required for transcription initiation. The consensus sequence (q.v.) is TATAAT.

Primary transcript The initial, and often very large, product of transcription of a eukaryotic gene. Subjected to processing to produce the mature mRNA molecule.

Primer extension Synthesis of a copy of a nucleic acid from a primer. Used in labelling DNA and in determining the start site of transcription.

Probe A labelled molecule used in hybridisation procedures.

Proinsulin Precursor of insulin that includes an extra polypeptide sequence that is cleaved to generate the active insulin molecule.

Prokaryotic The property of lacking a membrane-bound nucleus, e.g. bacteria such as *E. coli.*

Promoter DNA sequence(s) lying upstream from a gene, to which RNA polymerase binds.

Pronucleus One of the nuclei in a fertilised egg prior to fusion of the gametes.

Prophage A bacteriophage maintained in the lysogenic state in a cell.

Protease Enzyme that hydrolyses polypeptides.

Protein A condensation (dehydration) heteropolymer composed of amino acid residues linked together by peptide bonds to give a polypeptide.

Proteome Refers to the population of proteins produced by a cell. Cf. *genome*, *transcriptome*.

Protoplast A cell from which the cell wall has been removed.

Prototroph A cell that can grow in an unsupplemented growth medium.

Purine A double-ring nitrogenous base such as adenine and guanine.

Pyrimidine A single-ring nitrogenous base such as cytosine, thymine and uracil.

Random amplified polymorphic DNA PCR-based method of DNA profiling that involved amplification of sequences using random primers. Generates a type of genetic fingerprint that can be used to identify individuals.

RAPD See *random amplified polymorphic DNA*.

Reading frame The pattern of triplet codon sequences in a gene. There are three reading frames, depending on which nucleotide is the start point. Insertion and deletion mutations can disrupt the reading frame and have serious consequences, as often the entire coding sequence becomes nonsense after the point of mutation.

Recessive An allele where the expression is masked in the phenotype in heterozygous individuals. Cf. *dominant*.

Recombinant DNA A DNA molecule made up of sequences that are not normally joined together.

Recombination frequency mapping Method of genetic mapping that uses the number of crossover events that occur during meiosis to estimate the distance between genes. Cf. *physical mapping*.

Regulatory gene A gene that exerts its effect by controlling the expression of another gene.

Renaturation kinetics Method of analysing the complexity of genomes by studying the patterns obtained when DNA is denatured and allowed to renature.

Repetitive sequence A sequence that is repeated a number of times in the genome.

Replacement vector A bacteriophage vector in which the cloning sites are arranged in pairs, so that the section of the genome between these sites can be replaced with insert DNA.

Replication Copying the genetic material during the cell cycle. Also refers to the synthesis of new phage DNA during phage multiplication.

Replicon A piece of DNA carrying an origin of replication.

Restriction enzyme An endonuclease that cuts DNA at sites defined by its recognition sequence.

Restriction fragment A piece of DNA produced by digestion with a restriction enzyme.

Restriction fragment length polymorphism (RFLP) A variation in the locations of restriction sites bounding a particular region of DNA, such that the fragment defined by the restriction sites may be of different lengths in different individuals.

Restriction mapping Technique used to determine the location of restriction sites in a DNA molecule.

Retrovirus A virus that has an RNA genome that is copied into DNA during the infection.

Reverse transcriptase An RNA-dependent DNA polymerase found in retroviruses, used *in vitro* for the synthesis of cDNA.

Ribonuclease (RNase) An enzyme that hydrolyses RNA.

Ribonucleic acid (RNA) A condensation heteropolymer composed of ribonucleotides.

Ribosomal RNA (rRNA) RNA that is part of the structure of ribosomes.

Ribosome The 'jig' that is the site of protein synthesis. Composed of rRNA and proteins.

Ribosome-binding site A region on an mRNA molecule that is involved in the binding of ribosomes during translation.

RNA processing The formation of functional RNA from a primary transcript (q.v.). In mRNA production this involves removal of introns, addition of a 5′ cap and polyadenylation.

S_1 mapping Technique for determining the start point of transcription.

S_1 nuclease An enzyme that hydrolyses (degrades) single-stranded DNA.

Saccharomyces cerevisiae Unicellular yeast (baker's yeast) that is extensively used as a model microbial eukaryote in molecular studies. Also used in the biotechnology industry for a range of applications, as well as in brewing and bread-making.

Screening Identification of a clone in a genomic or cDNA library (q.v.) by using a method that discriminates between different clones.

SCIDS Severe combined immunodeficiency syndrome, a condition that results from a defective enzyme (adenosine deaminase, q.v.).

Scintillation counter A machine for determining the amount of radioactivity in a sample.

Selection Exploitation of the genetics of a recombinant organism to enable desirable, recombinant genomes to be selected over non-recombinants during growth.

Sequence tagged site Refers to a DNA sequence that is unique in the genome and which can be used in mapping studies. Usually identified by PCR amplification.

Sex chromosome The non-autosomal X and Y chromosomes in humans that determine the sex of the individual. Males are XY, females XX.

Sex-linked Refers to pattern of inheritance where the allele is located on a sex chromosome. Cf. *X-linked.*

Shine–Dalgarno sequence See *ribosome-binding site.*

Single-locus probe Probe used in DNA fingerprinting that identifies a single sequence in the genome. Diploid organisms therefore usually show two bands in a fingerprint, one allelic variant from each parent.

Single nucleotide polymorphism Polymorphic pattern at a single base, essentially the smallest polymorphic unit that can be identified.

Site-directed mutagenesis See *oligonucleotide-directed mutagenesis.*

SNP See *single nucleotide polymorphism.*

Somatic cell Body cell, as opposed to germ-line cell.

Somatotropin Growth hormone, see also *bovine somatotropin.*

Southern blotting Method for transferring DNA fragments onto a membrane for detection of specific sequences by hybridisation.

Specific activity The amount of radioactivity per unit material, e.g. a labelled probe might have a specific activity of 10^6 counts/minute per microgram. Also used to quantify the activity of an enzyme.

Sperm The mature male gamete. Cf. *ovum*, *ooctye.*

Sticky ends See *cohesive ends.*

Structural gene A gene that encodes a protein product.

STS See *sequence tagged site.*

Stuffer fragment The section in a replacement vector (q.v.) that is removed and replaced with insert DNA. See *polystuffer.*

Substitution vector See *replacement vector.*

Tandem repeat A repeat composed of an array of sequences repeated contiguously in the same orientation.

***Taq* polymerase** Thermostable DNA polymerase from the themophilic bacterium *Thermus aquaticus.* Used in the polymerase chain reaction (q.v.).

TATA box Sequence found in eukaryotic promoters. Also known as the *Hogness* box, it is similar to the Pribnow box (q.v.) found in prokaryotes, and has the consensus sequence TATAAAT.

T-DNA Region of Ti plasmid of *Agrobacterium tumefaciens* that can be used to deliver recombinant DNA into the plant cell genome.

Terminal transferase An enzyme that adds nucleotide residues to the 3′ terminus of an oligo- or poly-nucleotide.

Temperate Refers to bacteriophages that can undergo lysogenic infection of the host cell.

Tetracycline (Tc) An commonly used antibiotic.

Thermal cycler Heating/cooling system for PCR applications. Enables denaturation, primer binding and extension cycles to be programmed and automated.

Thermus aquaticus Thermophilic bacterium from which *Taq* polymerase (q.v.) is purified. Other bacteria from this genus include *Thermus flavus* and *Thermus thermophilus.*

Thymine (T) Nitrogenous base found in DNA only.

Ti-plasmid Plasmid of *Agrobacterium tumefaciens* that causes crown gall disease (q.v.).

Tissue plasminogen activator (TPA) A protease that occurs naturally, and functions in breaking down blood clots. Acts on an inactive precursor (plasminogen), which is converted to the active form (plasmin). This attacks the clot by breaking up fibrin, the protein involved in clot formation.

Totipotent A cell that can give rise to all cell types in an organism. Totipotency has been demonstrated by cloning carrots from somatic cells, and by nuclear transfer experiments in animals.

TPA See *tissue plasminogen activator.*

Traitor technology See *genetic trait control technology.*

***Trans*-acting element** A genetic element that can exert its effect without having to be on the same molecule as a target sequence. Usually such an element encodes a protein product (perhaps an enzyme or a regulatory protein) that can diffuse to the site of action.

Transcription (T_C) The synthesis of RNA from a DNA template.

Transcriptional unit The DNA sequence that encodes the RNA molecule, i.e. from the transcription start site to the stop site.

Transcriptome The population of RNA molecules (usually mRNAs) that is expressed by a particular cell type. Cf. *genome, proteome.*

Transfection Introduction of purified phage or virus DNA into cells.

Transfer RNA (tRNA) A small RNA (~75–85 bases) that carries the anticodon and the amino acid residue required for protein synthesis.

Transformant A cell that has been transformed by exogenous DNA.

Transformation The process of introducing DNA (usually plasmid DNA) into cells. Also used to describe the change in growth characteristics when a cell becomes cancerous.

Transgene The target gene involved in the generation of a transgenic (q.v.) organism.

Transgenic An organism that carries DNA sequences that it would not normally have in its genome.

Translation (T_L) The synthesis of protein from an mRNA template.

Transposable element A genetic element that carries the information that allows it to integrate at various sites in the genome. Transposable elements are sometimes called 'jumping genes'.

Trisomy Aneuploid (q.v.) condition where an extra chromosome is present. Common example is the trisomy-21 condition that causes Down syndrome.

Uracil (U) Nitrogenous base found in RNA only.

Variable number tandem repeat (VNTR) Repetitive DNA composed of a number of copies of a short sequence, involved in the generation of polymorphic loci that are useful in genetic fingerprinting. Also known as hypervariable regions. See also *minisatellite* and *microsatellite DNA.*

Vector A DNA molecule that is capable of replication in a host organism, and can act as a carrier molecule for the construction of recombinant DNA.

Virulent Refers to bacteriophages that cause lysis of the host cell.

Virus An infectious agent that cannot replicate without a host cell.

VNTR See *variable number tandem repeat.*

Western blotting Transfer of electrophoretically separated proteins onto a membrane for probing with antibody.

Xenotransplantation The use of tissues or organs from a non-human source for transplantation.

X-gal 5-Bromo-4-chloro-3-indolyl-β-D-galactopyranoside: a chromogenic substrate for β-galactosidase; on cleavage it yields a blue-coloured product.

X-linked Pattern of inheritance where the allele is located on the X-chromosome. In humans, this can result in males expressing recessive characters that would normally be masked in an autosomal heterozygote.

YAC Yeast artificial chromosome, a vector for cloning very large pieces of DNA in yeast.

Zygote Single-celled product of the fusion of a male and a female gamete (q.v.). Develops into an embryo by successive mitotic divisions.

Index

Page numbers followed by F, T and CM refer to Figures, Tables and Concept Maps, respectively.